Introduction to Microwave Electronics

T. C. Edwards

Lecturer in Electronics, Royal Military College of Science, Shrivenham

Edward Arnold

© T. C. Edwards 1984

First published 1984 by
Edward Arnold (Publishers) Ltd
41 Bedford Square, London WC1B 3DQ

Edward Arnold
3 East Read Street, Baltimore
MD21202, USA

Edward Arnold (Australia) Ltd.,
80 Waverley Road, Caulfield East,
Victoria 3145, Australia

Reprinted 1985, 1986

ISBN: 0 7131 3495 X

Printed in Great Britain by
Thomson Litho Ltd, East Kilbride, Scotland

Preface

This is an introductory book on the significant topic of *microwave electronics*. Satellite communications systems and modern radars, civil and military, demand a critical appreciation of the wide range of microwave devices and circuits which are available to the systems designer. This book provides a suitable treatment for this purpose, with the bare minimum of mathematics.

Most microwave technology now depends upon semiconductor devices and the first four chapters here are devoted to such devices and their applications; Gunn, IMPATT and TRAPPATT diodes, microwave transistors (concentrating on the gallium arsenide MESFET), and reflection amplifiers. At higher power levels microwave electron tubes are required and Chapter 5 deals with these: from the multi-cavity klystron, travelling-wave tube, and crossed-field tubes, to the gyrotron and reflex triode. In Chapter 6 various types of amplifier are compared and discussed and in the final chapter (7) microwave passive diodes are described. A selection of problems relating to microwave devices and their applications concludes the main part of the book.

This material has been developed from courses taught by the author both in the UK and in Melbourne, Australia, in particular the microwave electronics content of the MSc in Guided Weapons at Shrivenham and to some extent work covered within the undergraduate degree. It is felt that the text will also be appropriate for many other students on advanced systems-oriented courses, who require a background understanding of microwave electronics, but without the theoretical depth demanded of a full microwave study.

Students at Universities, Polytechnics, and other institutions should find this approach to the topic attractive and helpful and it should fill an introductory role for those wishing to specialize in microwaves.

Practising engineers in industry should also find the book useful and refreshing.

Acknowledgements

The author would like to thank Dr F J M Farley, previous Dean and Professor F R Hartley, Acting Dean, Royal Military College of Science (RMCS), Shrivenham, for permission to publish this text, and also Professor C J Harris, Head of Department of Electrical and Electronic Engineering, RMCS, for provision of facilities for the preparation of the manuscript. The Plessey Co. Ltd and Ferranti plc are thanked for their permissions to use photographs for the front cover, a gallium arsenide microchip and a TWT amplifier respectively. A special thanks is also extended to my students whose close attention to detail uncovered several initial errors and omissions, to Karen Pankhurst and Angela Hare for their patient and excellent typing, and to Sheila Prescott for producing the precise drawings.

TCE
1983

Contents

Preface   **iii**

Introduction to Microwave Engineering   **vii**
1 Diodes and Low-Power Oscillators   **1**
  1.1   The range of two-terminal microwave oscillating devices   1
  1.2   A qualitative description of Gunn domain formation   2
  1.3   Gunn oscillator circuits   6
  1.4   Power output and efficiency   7
  1.5   Phase-lock microwave sources (using a PLL)   7

2 Avalanche Transit-Time Devices   **10**
  2.1   Introduction   10
  2.2   The avalanche process   11
  2.3   The TRAPATT oscillator   13

3 Microwave Transistors   **17**
  3.1   Introduction   17
  3.2   Microwave bipolar transistors   17
  3.3   Microwave field-effect transistors   17
  3.4   Noise minimization   19
  3.5   Summary of solid-state devices' power/frequency
      performance   19
  3.6   Power and efficiency: some typical data   20
  3.7   Monolithic microwave integrated circuits   20

4 Reflection Amplifiers, and Digital Microwave Techniques   **21**
  4.1   Introduction: operating principles   21
  4.2   A resumé of digital modulation techniques   22
  4.3   A series-type QPSK modulator   23

5 Microwave Electron Tubes   **25**
  5.1   The planar triode   25
  5.2   The klystron   26
  5.3   The reflex-klystron oscillator   27
  5.4   The travelling-wave tube (TWT)   27
  5.5   The O-type backward-wave oscillator (BWO)   30
  5.6   The M-type backward-wave oscillator (or 'Carcinotron')   30
  5.7   Crossed-field amplifiers (CFA or, more generally, crossed-
      field tubes CFT)   30
  5.8   The magnetron   31

5.9   Summary of some significant tubes   33
5.10 The gyrotron and the reflex triode   34

6   Amplifiers: A General Comparative Discussion   36
6.1   Introduction   36
6.2   Selection of a suitable type of amplifier   36
6.3   Noise performance comparisons   37
6.4   A brief description of the parametric amplifier   39
6.5   Summary of low-noise amplifier trends   41
6.6   A comparative summary of microwave amplifiers   41
6.7   Noise and power-gain trade-offs in 'front-end' amplifiers
      (receivers)   41
6.8   Microwave solid-state power amplifiers: some basic
      aspects   42
6.9   Some special terms used in the context of power
      amplifiers   43

7   Microwave Passive Diodes   45
7.1   Introduction: detector performance criteria   45
7.2   The backward diode   47
7.3   The Schottky-barrier (or 'hot carrier') diode   48
7.4   Point-contact diodes   51
7.5   PIN diodes   52
7.6   Step-recovery diodes (SRD) (or 'snap-off' diodes)   56
7.7   Microwave varactor diodes   61

Problems   64

Answers   70

Suggestions for further reading   71

Appendices   72

1   Decibels (dB): Voltage and Power Ratios   72
2   Radio-Frequency Band Designations   74

Index   75

Introduction to Microwave Engineering

The term 'microwave' is actually rather ambiguous, although in exceedingly widespread use. However, the microwave bands might reasonably be said to extend from high u.h.f. through s.h.f. and into e.h.f. (see Appendix 2). Signals at frequencies from around 1 GHz to at least 100 GHz encompass the bands in question. Below 1 GHz the spectrum has allocations for services such as u.h.f. t.v., advanced mobile radios and the like. Fairly well-established technologies based upon coaxial lines and lumped (L, C, R) components are available for use in these areas. Above 1 GHz and through millimetric-wavelength bands, design tends to be considerably more difficult and the product more expensive.

Important systems which operate at microwave frequencies include the majority of radars, satellite communication systems, terrestrial line-of-sight links and tropospheric scatter links. This book describes microwave electronic components and sub-systems which are encountered during the development of systems such as those mentioned, and in their use.

Microwave circuits e.g. couplers, amplifiers, oscillators, detectors etc. were intensively developed during the World War II period to meet the pressing military demands of radar, in particular. Bearing in mind that the transistor was not invented until 1948, the only semiconductor element used in early 1940's-style microwave circuits was the crystal detector diode. Waveguide pipes predominated in the transmission technology, even where power levels were relatively low, and as a result of this microwave engineers gained the nickname 'electronic plumbers'. Vacuum tubes such as magnetrons, klystrons and travelling wave tubes (TWTs) served signal generation and amplification functions. At kW levels and above, where solid state cannot yet reach, the vacuum tubes remain in use and have been the subject of considerable developments for microwave purposes.

Since the late 1960's the situation has changed quite dramatically. Semiconductor elements such as Gunn and avalanche diodes are available to generate or (for special requirements) amplify power. Furthermore, both bipolar and field-effect transistors have become available which have useful power gain at microwave frequencies. Also in the 1960's it became apparent that hybrid micro-circuit technology could readily be adapted to provide microwave integrated circuits (MIC).

Gallium arsenide monolithic microwave integrated circuits, and related high-speed digital circuits, are now under intensive development.

1

Diodes and Low-Power Oscillators

1.1 The Range of Two-Terminal Microwave Oscillating Devices

It is now feasible to design certain microwave transistor oscillators to operate up to about 40 GHz, but a range of two-terminal devices also exists which provide microwave oscillation, and several situations demand these. Since many microwave systems are required to operate at frequencies in the millimetric ranges there exists a clear need for solid-state devices to provide millimetric oscillators—as well as other functions such as amplification via reflection amplifier configurations. Several special devices, operating on different principles, have been developed to fulfil these pressing needs. It is important to emphasize that the devices which will be described can also very readily be designed to operate at the lower microwave frequencies (including in some cases u.h.f.). These special devices tend to be used, at 'transistor frequencies', for one or more of the following reasons:

(i)   Cost: transistors are often much more expensive.

(ii)   Technological compatibility: transistor circuits are invariably exceedingly difficult and inflexible to design in a waveguide or even coaxial format. The special devices are 'naturals', being two-terminal, for waveguide or coaxial environments where this is warranted. (They are also certainly available in MIC packages).

(iii) Low-noise (e.g. Gunn or BARITT) or relatively high-power (e.g. IMPATT, TRAPATT) – in both cases several two-terminal solid-state devices are available with a microwave performance surpassed only by special valves or 'vacuum-tubes'.

IMPATT, TRAPATT and BARITT diodes are summarized in Section 2.1.

The solid-state microwave oscillating devices fall very broadly into two classes: one which makes use of bulk semiconductor behaviour (i.e. no junction effects used), and a second class which specifically uses junction effects. It is worth briefly listing the general types of device:

(a) Tunnel Diodes

These devices have been available for many years now. They consist of heavily doped PN junction diodes in which the electron tunnelling phenomenon occurs. As a result of this a negative slope resistance is manifest over a restricted range of the forward I/V characteristic. Negative-resistance oscillators may therefore be designed, albeit with some difficulty and often considerable lack of control! The main application area has turned out to be negative-resistance reflection amplifiers—several Intelsat satellite amplifiers used Tunnel Diode Amplifiers—or TDA's. This application emerged due to the tolerably low-noise performance, bearing in mind that suitable transistor amplifiers were not available sufficiently early (1965-1971 era).

(b) Transferred-Electron Oscillators (TEOs)

TEOs are sometimes considered to be just another name for Gunn diodes. This is really a gross over-simplification, since Limited Space charge Accumulation diodes (LSA diodes) are also TEOs and are quite distinct from Gunn diodes, which are described very shortly.

(c) 'Transit-Time' Devices

This class of device is characterized by a very definite dependence of oscillation conditions upon the transit-time for a bunch of charges to move through a specific region of the device. The 'charge cluster' can be produced by avalanche multiplication (IMPATT), injected charge and punch through (BARITT) or a technique depending upon avalanche and a 'trapped plasma' (TRAPATT). These devices are described in Chapter 2.

The devices do not really compete with each other for applications to any great extent. Instead, various devices tend to suit specific applications where others fall short of the demands. For example, Gunn diodes have a well-established reputation in C.W. short-range doppler radars, and as local oscillators in microwave receivers. They are low-to-moderate power and tolerably low-noise devices. On the other hand, the contrastingly noisy IMPATTs can provide several watts of continuous-working C.W. power, which makes them useful in applications such as final amplifiers for microwave cross-country link transmitters.

The spectrum of applications shown as Fig. 1 gives an idea of the setting of microwave solid-state devices.

1.2 A Qualitative Description of Gunn Domain Formation

It has been established that gallium arsenide (GaAs) exhibits a very non-linear velocity–field characteristic, including a region of negative resistance. The next step is to develop a qualitative description leading to domain formation—conceptually—in a Gunn diode.

Inhomogeneities in the bulk material produce local variations in the electric field through the active layer. In particular, at the cathode where electrons enter the material, crystal damage almost invariably leads to local fields larger than the average field in the material. Free electrons in this (near cathode) region will experience a field above the threshold level before any others. (In GaAs there exists a minimum electric field required for the initiation of the transferred-electron effect, and this we term the threshold field.) Also, the material possesses an energy band structure that is quite

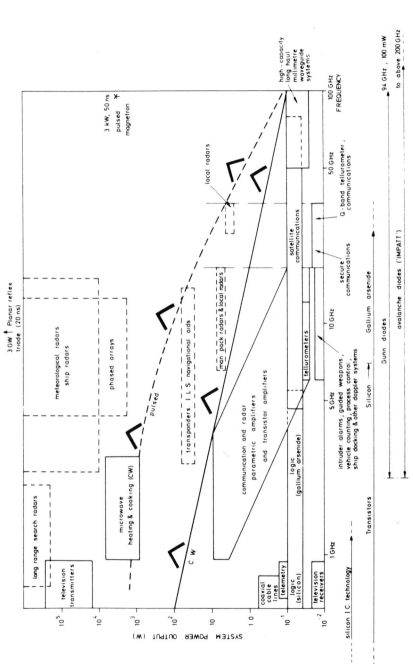

Fig. 1 Spectrum of applications for a wide range of microwave systems (with acknowledgement to Dr F W Myers, Plessey Co. Ltd for much of the detail here).

unlike the structure of a semiconductor such as silicon. At fairly low electron velocities a so-called 'lower conduction band' applies and the mobility is high (several thousand). As electron velocities increase they reach a new condition that is associated with a so-called 'upper conduction band', for which the mobility is about an order of magnitude lower than for the lower band. Therefore, once electrons are travelling sufficiently fast to reach the upper conduction band they are rapidly decelerated as a consequence of the low mobility.

1. SCHEMATIC SKETCH OF GUNN DIODE

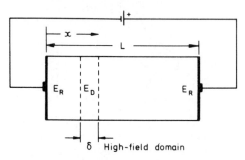

2. DOMAIN PROFILE IN GUNN DIODE

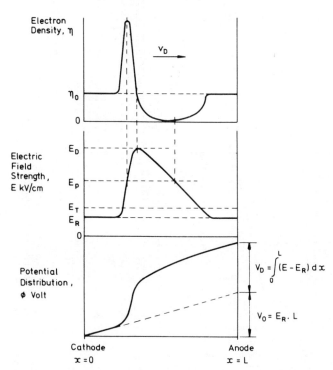

Fig. 2   The Gunn diode: schematic sketch of the device and the profile of a travelling domain.

The situation is analogous to a traffic queuing situation when one vehicle suddenly slows down. Electrons ahead of the slowed-down bunch will continue at their original high velocity and a depletion layer will grow. Electrons behind will catch up the slow ones until they also transfer to the low-mobility state. An accumulation 'spike' of electrons is thus formed. The resulting picture of the charge distribution through the sample together with electron density, potential and electric field is shown in Fig. 2.

This situation is correct once a fully depleted stable accumulation-depletion domain has formed.

A built-in field profile is set up across the domain due to the charge distribution shown in the diagram. For every value of electric field between E_R and E_D there are two possible values of charge density n: one being $<n_0$ and the other $>n_0$.

It is sensible to get the physical dimensions and electrical orders-of-magnitude into perspective. The active semiconductor layer is usually about 10μm thick ($L \simeq 10\mu$m) and consists of n-type epitaxially-grown GaAs. This layer is grown on top of a fairly heavily doped n + layer forming an ohmic contact. The other end of the device also has an n + layer to form the opposite contact. One end will be the cathode and the other will be the anode. It is important to appreciate that the operation of the device depends entirely on the bulk properties of the material—no junction effects are involved at all.

When a voltage is applied across the device, the behaviour is initially ohmic, it appears that a GaAs resistor has just been connected across the supply. However, once the electric field across the active layer exceeds about 3.2 kVcm^{-1} (only about 3.2 V for 10 μm region), the current falls sharply to a lower value and regular current spikes at microwave frequency appear in an external resistive circuit. The domain-formation mechanism, already described, is responsible for these current spikes. The frequency of the spikes is inversely proportional to the thickness of the active layer and a 10 μm device yields a frequency of about 10 GHz.

An engineer wishing to apply this type of device will need to know the answers to several practical questions, for instance:

(i)   What determines possible frequencies of oscillation?
(ii)   What determines the required supply voltage?
(iii)   How much r.f. power may be delivered, and at what conversion efficiency?

(iv)   Is the device suitable, in one form or another, as an amplifier active element?

and

(vi)   Other items such as: packages, availability, cost, and the vital question of reliability.

It will be found that, although the frequency is mainly determined by v_D/L, some modification occurs especially with the so called 'quenched-domain' mode.

$$\textit{Natural} \text{ frequency for diode: } f_o = \frac{v_D}{L}$$

By investigating the Gunn mechanism in more depth a quantitative verification of a mobile domain can be obtained. Also the velocity–field characterisic can be more accurately defined, leading towards an answer to question (ii)—the d.c. supply voltage.

These analyses are the province of the specialist device department scientist.

1.3 Gunn Oscillator Circuits

Many practical Gunn oscillators are made available in waveguide form. There are mounting, tuning and inherent frequency stability advantages to be enjoyed with the waveguide resonator. The practical arrangement and equivalent circuit are shown in Fig. 3.

As shown in Fig. 3(a), the d.c. bias supply power is fed to the Gunn diode via a decoupled terminal. Further r.f. decoupling (or even full low-pass filtering) may be necessary to ensure that the bias supply has a relatively high r.f. impedance. The remainder of the support structure consists of a cylindrical post, height h, across the waveguide. A waveguide resonator is thus formed between this structure, the sliding short-circuit plunger, and the iris output coupler. It would also be possible, although much less convenient, to use either a waveguide junction tuner or a set of tuning screws between the device and the load to which the r.f. output is delivered. In the arrangement of Fig. 3 the iris serves to ensure that the oscillation conditions are fairly insensitive to wide variations of load impedance.

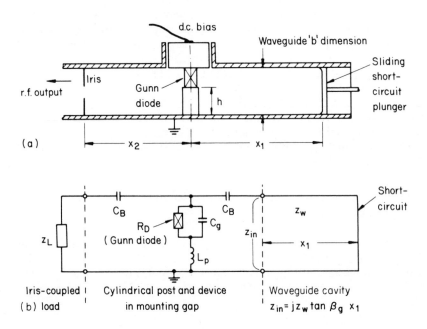

Fig. 3   A Gunn oscillator based on a rectangular waveguide cavity.

1.4 Power Output and Efficiency

Practical Gunn devices may have a C.W. output power as low as a few milliwatts or as high as a few watts. Under pulse conditions LSA devices deliver 100 to 200 W, providing a fairly low duty cycle and narrow pulse width are employed.

Many C.W. devices are specified at some tens of milliwatts of r.f. output and their applications range from C.W. radar to microwave receiver local oscillators. There are some strict limitations regarding power and frequency, and these are described shortly.

Efficiency is of little interest with very low-power devices, but it becomes important for devices which deliver r.f. power from a fraction of a watt upwards. In these cases the thermal dissipation must be known. Good heat-sinking for the device is essential in nearly all cases.

In practice Gunn oscillator efficiencies (η) usually lie between the following extremes:

$$0.2 \leq \eta \leq 20\%$$

A 2 W device operating at 10 GHz typically has an efficiency of just over 9%. About 22 W of d.c. supply power is therefore required for 2 W of r.f. output (high for a Gunn diode). Cooling is therefore a considerable problem.

Many low-power C.W. Gunn oscillators, which are very popular, have efficiencies in the region of only 0.5%.

1.5 Phase-Lock Microwave Sources (using a PLL)

On fundamental grounds, and from a practical manufacturing viewpoint, a phase-lock microwave source incorporating a phase-lock loop with a phase detector (PLL) is an excellent method of designing a high-grade microwave source. There is a need for such high-grade sources in some civilian and several military systems such as 'identification', friend or foe' (IFF).

Furthermore, a PLL-based microwave source is very conveniently frequency modulated with a wide modulation bandwidth being easily obtained. Also automatic programming for the synthesis of microwave signals is readily achieved.

An example of an arrangement for a phase-lock microwave source is shown in Fig. 4.

Under normal steady-state, or near steady-state, conditions only the main phase-lock loop (PLL) is operative and the microwave oscillator frequency is brought into lock with the harmonic generator output frequency f_n via the conventional PLL dynamics. With these conditions, once lock has been established, the phase detector (actually realized using a balanced microwave mixer) produces a small steady output voltage v_ϵ which is amplified by the phase-lock amplifier and fed to the frequency-controller of the microwave oscillator. In such a sensitive system, this frequency-controller need not be a separate (e.g. varactor) element: the bias-sensitivity of a Gunn diode oscillator, for instance, may well be sufficient.

Following initial switch-on, it may not be possible for the locking process

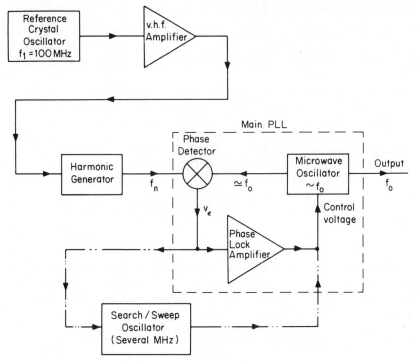

Fig. 4   A typical phase-lock oscillator arrangement.

to begin via the main PLL because the two frequencies f_o and f_n may differ too greatly. Then the output of the phase detector will remain saturated and no self-synchronizing transients can start. In order to overcome this problem a secondary loop consisting of a comparatively slow (MHz) search/sweep oscillator is introduced. The output from this is of sufficient amplitude to tune the frequency of the microwave oscillator over a wide enough range for it to coincide, at some time, with f_o.

Once this happens the phase detector output rapidly drops and causes the search/sweep oscillator to be switched off. After this the main PLL is the only operative loop. In fact, the search/sweep oscillator is only switched on under conditions when the phase-detector output has remained saturated for several microseconds. The crystal oscillator and the v.h.f. amplifier are both more or less conventional. It is usually necessary for the amplifier to deliver around 100 mW or so of power, to make up for considerable losses incurred in the harmonic generator. This harmonic generator might use varactor diodes or step-recovery diodes to provide the highly non-linear waveforms which are rich in harmonics. Whichever devices are used, a frequency multiplication factor of 40 would be required for 4 GHz and, with 100 mW input, only a few milliwatts of power will be applied to the phase detector.

It is useful, finally, to list the considerable specification advantages offered by phase-lock microwave sources; they have:

(i)   A very high rejection of harmonics generated by the reference. These are typically -125 dBc.

(ii)   The microwave output has the same long-term stability as the reference.

(iii)   The tunability is good since the units are electronically and mechanically flexible.

(iv)   The FM noise is the lowest obtainable from any currently conceivable microwave source (recollect that FM noise is the dominant noise in many microwave oscillators; especially Gunn devices).

(v)   Frequency Modulated signals can be produced, with a wide modulation bandwidth.

(vi)   These phase-lock sources can readily be automatically programmed as synthesizers.

Recent and current developments are resulting in SAW* (and BAW*)—controlled oscillators where the fundamental frequency is between 1 and 2 GHz typically. Thus, even where such sources are not actually used alone, they may replace the crystal reference oscillator and v.h.f. amplifier in a phase-lock source. For instance, a 10.2 GHz final source could be derived in the manner explained, but using a 1.7 GHz SAW reference oscillator and only a ×6 multiplier. In this way the amplifier is not required and several multiplication stages are saved. Also, the circuitry is likely to involve less volume than that required with a v.h.f. crystal.

*SAW—Surface Acoustic Wave.   BAW—Bulk Acoustic Wave.

2

Avalanche Transit-Time Devices

2.1 Introduction

This class of device has previously been briefly mentioned (page 00). All of these transit-time devices differ distinctly from transferred-electron devices in that they do not involve travelling domains. Instead, junction effects between differently doped semiconductor materials are responsible for their operation. There are three major types of device in this category which are listed here in an order indicating the *extent* of their applications:

(i) IMPATT diodes (IMPact Avalanche Transit-Time).

(ii) TRAPATT diodes (TRAPped (plasma) Avalanche Transit-Time), and

(iii) BARITT diodes (BARrier Injection Transit-Time).

IMPATT diodes are prominent in certain oscillator applications and in the moderate power stages of reflection amplifiers. Whilst they produce substantially greater power than Gunn diodes can achieve, this is heavily offset by the considerably greater noise. Thus, IMPATTs tend to complement rather than compete with Gunn devices in many C.W. and in a few pulsed applications.

TRAPATT diodes operate almost invariably in pulsed form, and they can deliver considerable microwave power. In fact, TRAPATTs are the leaders in this respect for many microwave systems applications and some hundreds of watts are obtainable for a sub-microsecond pulse, at a frequency of a few Gigahertz.

BARITT diodes are relatively new on the scene and possibly have a rather uncertain future. They are, it seems, the 'quietest' solid-state microwave oscillator diode available but, as might be expected, they also produce only a restricted power output—usually only a few milliwatts. In spite of this low power their low noise allows them to compete quite strongly with Gunn devices, for example in short range Doppler radar applications. Their main drawback will probably be the high supply voltage required (compared with Gunn).

Before going on to describe the avalanche process and the diode operation, it is worth briefly reflecting on the general developments leading to present-day devices.

In 1958, at Bell Telephone Laboratories, W T Read put forward the first proposals deducing that a properly structured diode, reverse-biased into breakdown, should give rise to negative resistance, i.e. the basic idea has been available for over two decades. It is only since about 1965 that suitable device structures have been directly developed for IMPATT diodes. These microwave devices can be made from silicon or gallium arsenide. Due mainly to the well-established technology and the better thermal behaviour the majority of IMPATT diodes are made of silicon.

The acronym 'IMPATT' stands for IMPact Avalanche Transit-Time, and this term gives some indication of the process involved in this diode. The avalanche process is described in the next section.

2.2 The Avalanche Process

To illustrate the meaning of avalanche breakdown, consider the depletion layer of a reverse-biased PN junction diode as shown in Fig. 5(a).

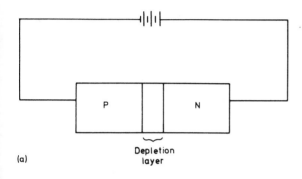

(a)

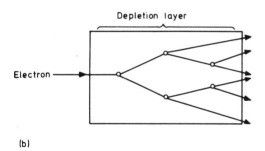

(b)

Fig. 5   Avalanche breakdown.

In Fig. 5(b) the history of one electron is depicted (holes are not drawn on this diagram—they multiply from right to left). Each dot indicates an ionizing collision, i.e. a collision of an electron with the lattice which is sufficiently violent to free a valence electron and deflect the incident electron. By the nature of this process, one hole–electron pair is generated by every ionizing collision. Thus the particle current carried by electrons increases from left to right, while the hole current increases from right to left. The sum of these two currents at any point in the diode is of course the total direct current supplied. A schematic view of the structure of an avalanche diode, and parameter distributions, is shown in Fig. 6.

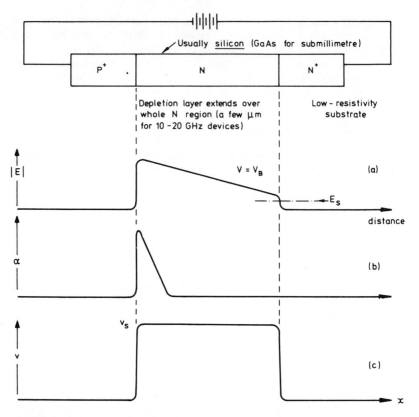

Fig. 6   IMPATT diode: electric field (a), ionization rate (b) and drift velocity (c) as approximate functions of distance through the device.

The dependence of ionization rate α on electric field E is so sensitive that, to a fair approximation, the avalanching can be considered to be concentrated in the left-hand end of the depletion layer. In this simple picture, it is assumed that there is no avalanching except in this narrow region close to the electric field peak.

The diode has been designed so that the depletion layer just reaches the N⁺ substrate at breakdown (i.e. it is just 'punched through'). This is to ensure that electrons and holes crossing the region do so at constant velocity V_s (approx. 10^7 cm s⁻¹). Thus, a bunch of electrons injected at one side of the depletion region would remain together, crossing it at constant velocity. In practice there is slight dispersion of the bunch due to diffusion, but it will be ignored here for the sake of simplicity.

To understand how the IMPATT mode can give a.c. negative resistance it is necessary to explain how the avalanche and drift processes can combine to give more than 90° phase shift between current and terminal voltage. The simplest case to illustrate is the one in which the phase shift is 180° i.e. the device is a pure negative resistance. (The only currents considered here will be particle currents; there is also a displacement current present which could

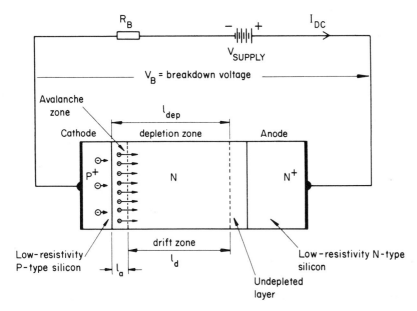

Fig. 7 Schematic sketch of an IMPATT diode biased just into reverse breakdown.

be allowed for—approximately—using the capacitance of the reverse-biased junction). The situation is indicated in Figs. 7 and 8.

The cavity in which the IMPATT diode will operate must have a net value of inductance to 'tune out' the device capacitance at the frequency of oscillation. This is shown, together with the load resistance, in Fig. 9.

From the usual oscillation condition $-Z_D = Z_L$, the frequency is clearly dependent upon the device capacitance and the resonator inductance. The value of the load resistance is very significant, since at least it determines whether oscillation or amplification will occur. If the circuit acts as an amplifier, R_L will not intersect the $-R_D$ versus I_D curve, and the diode-plus-load circuit will generally be connected to the appropriate part of a circulator to form a stable negative resistance amplifier.

The gain is then a function of $-R_D$ and line characteristic impedance Z_o. Since then $|R_D| \ll Z_o$, there is a considerable impedance transformation to be carried out for coaxial lines (50 ohms) or full-height waveguide (400 ohms). A ridged waveguide can be used for this impedance transformation.

2.3 The TRAPATT Oscillator (TRAPped (plasma) Avalanche Transit Time)

TRAPATT operation can be explained with little or no reference to the IMPATT mode, but TRAPATT circuits are easier to understand if this mode is approached as a limiting case of IMPATT operation.

Suppose, for example, that at first a typical IMPATT oscillator is considered, shown schematically in Fig. 10.

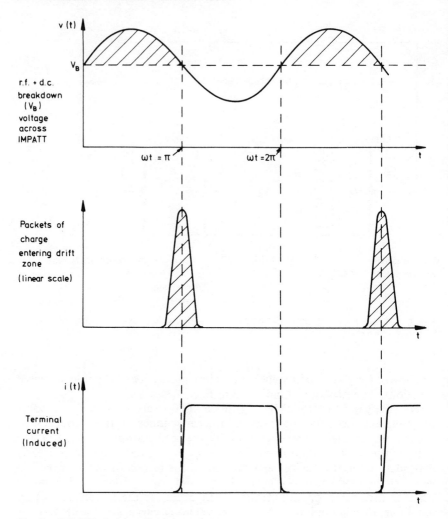

Fig. 8   IMPATT diode oscillator: waveforms.

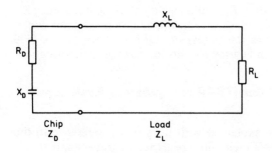

Fig. 9   IMPATT diode chip and load equivalent circuit.

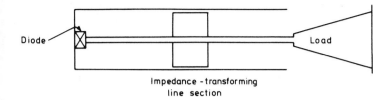

Fig. 10 An arrangement for a coaxial IMPATT circuit.

The diode is working at, say, 10 GHz at 5% efficiency as a normal IMPATT. Now, the rules of IMPATT operation are deliberately broken as follows:

(i) Increase the current density so that space charge effects depress the field below E_s for part of the cycle. (It may be necessary to use short pulses of current in order to avoid disastrous overheating effects).

(ii) Replace the load by a short-circuit, so that the diode is only loaded by the cavity losses. This will cause the voltage amplitude to grow until something limits it (usually device non-linearities)—again, E falls below E_s somewhere.

Three results may be anticipated as a consequence of this rather drastic action:

(i)   Because $E < E_s$ the transit time will increase and the frequency of operation will fall:

(ii)   Because of the large voltage modulation, the efficiency will be higher than a normal IMPATT mode.

(iii) No power will be supplied to anything outside!

The last problem is easily disposed of by using a low-pass filter instead of a short-circuit as indicated in Fig. 11.

Power can be extracted at the lower frequency, yet the IMPATT frequency circuit still sees an effective short-circuit because this frequency lies in the stop band of the filter. In practice it has been found that the original 10 GHz IMPATT at 5% efficiency can give 2 GHz at 25% or

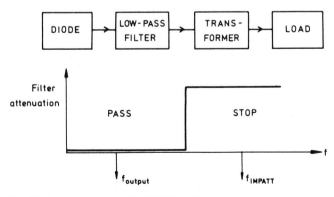

Fig. 11 Organization of a TRAPATT oscillator.

perhaps 1 GHz at 50% efficiency. The TRAPATT mode needs higher current densities than one normally meets with IMPATTs (thousands rather than hundreds of amps cm^{-2}) and is a low-frequency mode in the sense that a diode operated as a TRAPATT will operate at least a factor of 2 below its transit-time frequency—perhaps up to a factor of 10 lower.

In practice, a circuit devised by Evans of BTL enables the TRAPATT frequency to be controlled rather simply. This is shown schematically in Fig. 12.

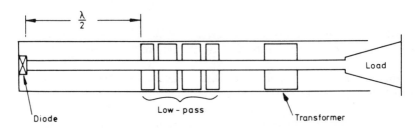

Fig. 12 Schematic diagram of the Evans TRAPATT circuit (λ is the wave-length at the TRAPATT frequency).

The low-pass filter can be considered to represent a short-circuit plane for the harmonics of the TRAPATT output frequency, and this can be used to trigger the avalanche wave before the IMPATT oscillations build up sufficiently. A steep drop in voltage causes a large-amplitude negative voltage pulse to be sent down the transmission line. When it reaches the low-pass filter it is reflected with substantially unchanged amplitude but 180° phase reversal. On reaching the diode again, it momentarily adds a large voltage (of the order of the breakdown voltage) to that already existing, with the result that the avalanche wave is launched. The condition for this is that the low-pass filter should be half a wavelength away from the diode at the TRAPATT frequency.

An interesting point arises here about the precise role played by the IMPATT oscillation in practical TRAPATT circuits. Evidently, once the first avalanche wave has been launched TRAPATT oscillations in the Evans circuit will be self-sustaining because the arrival of the reflected, inverted voltage transient will ensure that the next avalanche wave is launched, even in the absence of growing IMPATT oscillations meanwhile. It would appear, on this simple argument, that the role of the IMPATT in this circuit is simply to bring about the launching of the first avalanche wave; after that, the circuit takes over and sustains the TRAPATT mode.

As solid-state devices go, the r.f. (microwave) power obtainable from a TRAPATT device and circuit is relatively high. In amplifier form 70 W is obtainable, pulsed for 0.1 μs at 0.1% duty cycle, with a carrier frequency of 2.25 GHz.

These devices are unsuitable for frequencies above a few Gigahertz. It should be mentioned that, typically, four TRAPATT devices can be combined to produce over 200 W of pulsed power. Applications include wholly solid-state short range pulsed radars.

3

Microwave Transistors

3.1 Introduction

The main circuit features of microwave transistors will later be discussed in connection with microwave transistor amplifiers. In this section the fundamental device principles and limitations are described. As at lower frequencies, there are two general classes of device: bipolar transistors and field-effect transistors (FETs). Also, the most basic principles of operation are similar to those for lower-frequency devices. The major effort in designing microwave transistors goes into special techniques to raise the transition frequency f_T to as high a value as possible.

3.2 Microwave Bipolar Transistors

In spite of the great momentum obtained by gallium arsenide field-effect transistors, the 'glamour devices' of the current era, silicon bipolar transistors are still used in many systems operating at 'C-band' and below (less than about 8 GHz). Designers using silicon bipolar transistors—'bipolars' for short—often refer to them meaningfully as 'old reliable'.

Bipolar microwave transistors are nearly always constructed using silicon, which is principally because of the inherent stability of silicon monoxide (SiO) which is formed as an insulating and protective layer. Patterns with dimensions around 1 μm have to be etched into the oxide and the extremely small tolerances required are not obtainable with this kind of process in gallium arsenide. No further device details will be included here regarding bipolar transistors, due to the greater significance of field-effect devices.

3.3 Microwave Field-Effect Transistors

These devices have overtaken bipolar transistors, especially for frequencies above about 8 GHz. In particular, field-effect transistors are the only types of transistors which can provide useful gain with low-noise at frequencies through J-band i.e. up to and beyond 20 GHz (current research: to 40 GHz).

Currently, the vast majority of applications are filled by the special FET devices known as MESFETs. This term is short for MEtal Semiconductor Field-Effect Transistor, and the first three letters emphasize the importance of a metal semiconductor contact for the gate.

The earliest developments took place in 1969, principally by a researcher called Middlehoek. This earliest work was based upon silicon, but the photo-lithographic techniques were good even by today's standards since 1 μm gates were made. Even so, the use of silicon was the main reason for the rather low frequencies—the maximum frequency of oscillation for Middelhoek's device was 12 GHz. Technology since 1969 has concentrated strongly on gallium arsenide, for a number of reasons. With 1 μm gate lengths GaAs MESFETs achieve a maximum frequency of oscillation of about 50 GHz. Some devices are now being made with 0.5 μm gate lengths and, since electrons possess a higher mobility than holes, devices are generally made N-channel (see Fig. 13). Improvements due to GaAs technology are mainly as follows:

(i) (electron mobility in GaAs) = 6 × (electron mobility in Si).

(ii) (saturation velocity in GaAs) = 2 × (saturation velocity in Si).

These factors mean that, compared with silicon, parasitic resistances can be made smaller, mutual conductance (g_m) can be larger and the high-field electron transit-time is shorter. Also,

(iii) The active layer can readily be grown on a semi-insulating GaAs substrate (resistivity $> 10^7 \Omega - cm$).

This allows the parasitic capacitance associated with the gate bonding pad to be almost totally eliminated by correctly positioning the pad on the substrate.

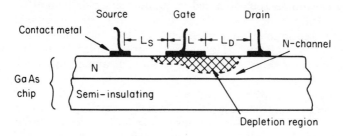

Fig. 13 Cross-section of a Schottky-gate GaAs MESFET.

Figure 13 shows a single (Schottky) gate MESFET, of a low-noise type, with its depletion region in saturation. There is little difference between the basic control behaviour for this device and that for devices operating at much lower frequencies. A negative bias potential is applied to the gate electrode and the magnitude of this bias then controls the effective width of the depletion region, and hence the amount of current flowing through the N-channel.

Although this discussion concentrates on the single-gate type of device, dual-gate devices are also available. These are useful in applications such as modulators, mixers and a.g.c. (automatic gain control).

Power MESFETs are also becoming available. These are constructed slightly differently in that deep, highly-doped (N-type) diffusions extend downwards from both source and drain.

Typically, 2 to 3 W is obtainable at 12 GHz.

For all these devices, the maximum frequency of oscillation (f_{max}) is inversely proportional to the gate length (L). Therefore a major aim is to keep L as short as practicable and some devices now have sub-micron gate dimensions.

3.4 Noise Minimization

Significant applications occur in low-noise circuits for either the bipolar transistor or, more often, the MESFET and various specific applications are described in Chapter 6.

From the device viewpoint it is clearly important to choose a suitably low-noise specification initially. In the case of a MESFET the d.c. drain current is then set at an optimum value for minimum noise figure. The overall circuit noise figure can then almost always be reduced still further by incorporating a noise-matching microstrip circuit between the source and the device.

3.5 Summary of Solid-State Devices' Power/Frequency Performance

This is illustrated in Fig. 14, where various diode devices are shown as well as transistors.

For the GaAs TE (Transferred Electron) devices, the maximum power−resistance product is close to 5×10^5 watt-ohms. Therefore the

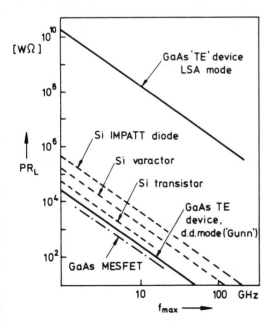

Fig. 14 (r.f. power).(load resistance) product to a base of frequency for several microwave semiconductor devices. (TE−transferred electron (Gunn); d.d. −delayed domain).

minimum possible frequency is somewhat less than 1 GHz.

Silicon IMPATT devices possess generally higher power–resistance products than Gunn devices or bipolar transistors, but there are drawbacks, especially noise, which restrict their applications.

The LSA devices clearly outperform all the other devices— easily—although difficulties are sometimes experienced in operating these devices. (Even then they are only suitable for submicrosecond pulse operation.)

3.6 Power and Efficiency: Some Typical Data

Gunn (C.W.): up to a few watts, $\eta < 20\%$ (max).
IMPATT (C.W.): up to 1 W at $\simeq$ 96 GHz, η = 5-10% (at about 10 GHz).
GaAs MESFET: 2 W at 12 GHz, $\eta \simeq$ 30%.

Other aspects regarding power and noise are dealt with in Chapter 6.

3.7 Monolithic microwave integrated circuits (MMICs)

The GaAs MESFET has, so far, been applied mainly in hybrid integrated circuits or MICs. In these circuits the transistor is usually embedded in a microstrip configuration that provides matching and d.c. supply filtering. These MICs will continue to be implemented in many systems.

Monolithic circuits, or MMICs, have been under intensive development for some time and are principally based upon a high-resistivity GaAs microchip. Unlike lower frequency microchips it is not feasible to incorporate large numbers of transistors within each analogue circuit. This means that, unfortunately, a substantial area of the chip must be used for lumped or distributed matching and supply filtering circuitry. Successful MMICs are however emerging from various research laboratories and are likely to make a significant impact in areas as diverse as direct broadcast satellite (DBS) receivers and phased-array radars, where high volume production is conceivable.

GaAs MESFET MMICs may be designed for power or, alternatively, low-noise operation. Power MMICs achieve over 1W C.W. to at least 8 GHz whereas low-noise circuits achieve about 3 dB noise figure with 25 dB gain to at least 11 GHz.

4

Reflection Amplifiers, and Digital Microwave Techniques

4.1 Introduction: Operating Principles

In microwave *transistor* amplifiers the device which is driven by the r.f. signal input, directly converts available d.c. power into r.f. signal output power. Although transistors are available which provide useful gain at frequencies of at least 20 GHz, microwave amplification is still required for communications and radar at frequencies well beyond 20 GHz—even up to hundreds of GHz. Furthermore, the maximum output power which can be obtained from microwave transistors is presently somewhat modest.

With a view to filling these, and other, gaps in systems requirements an alternative amplifying arrangement has been conceived and developed. In this alternative configuration a circulator is normally employed. It is not necessary to employ transistors in order to achieve amplification in such configurations. The increase in r.f. power takes place via reflection from an active two-terminal device which is connected to the 'intermediate' port of the circulator—between source and load. To emphasize the contrasts between transistor amplifiers and these reflection amplifiers the general arrangements are sketched in Fig. 15.

The circulator passes r.f. power with only a fraction of a dB loss from port-to-port in the direction of the arrow in Fig. 16. Some 20 dB (typically) of attenuation is offered in the opposite direction. (This is achieved by means of a magnetized ferrite disc). Thus, r.f. input power enters port 1, travels to port 2 with little loss of power, is reflected from the active two-

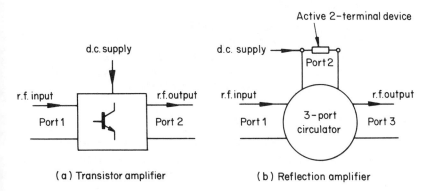

(a) Transistor amplifier　　　　(b) Reflection amplifier

Fig. 15 A comparison of the general arrangements for (a) transistor and (b) reflection amplifiers.

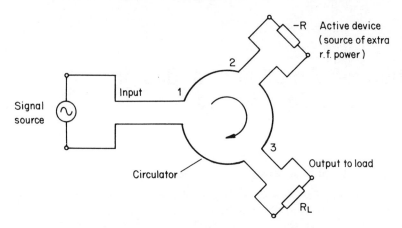

Fig. 16 Schematic diagram showing reflection amplifier operation.

terminal device at an increased level, and passes to the output on port 3 again with very little loss. In this way a net power gain is achieved.

For the moment, an abbreviated list is given of devices which can potentially provide the requisite negative resistance. These are:

(i) Tunnel diodes. These familiar devices have been with the electronic engineering fraternity for about two decades. Negative slope resistance is manifested because the tunnelling phenomenon produces a dip in the direct I/V characteristic.

Tunnel diode amplifiers (TDAs) have been used, for example, in existing satellite repeaters for the Intelsat (and other) satellite communication systems, particularly in the spacecraft's r.f. receivers.

(ii) Gunn diodes (or variants thereof), IMPATT diodes and BARITT diodes.

Circuits using these devices are becoming increasingly popular, especially where high gain and/or moderate power multi-stage microwave amplifiers are required.

(iii) Parametric circuits using pumped varactor diodes. These will be described briefly in Chapter 6.

Amplifiers based upon devices mentioned under (i) or (ii) tend to be more noisy than the 'quieter' microwave transistor amplifier designs, and certainly much noisier than parametric amplifiers. But they can provide considerable gain and power, and they can be designed to operate at much higher microwave frequencies than transistor amplifiers.

4.2 A Resumé of PCM Modulation Techniques

The use of phase-shift-keying (PSK) in time-division-multiple-access (TDMA) communication networks is a significant consideration for several microwave communication systems. Notable examples include:

(i) The forthcoming generations of satellite systems.
(ii) High-capacity terrestrial microwave links.

The modulation efficiency, in bits s^{-1} Hz^{-1}, is a strong function of the order applicable to the PSK used. Simple (2-level) PSK only offers an efficiency of 0.5 whereas, at the other extreme, 16-level PSK yields an efficiency of 4.0. However, the complexity of the technology required rises considerably with the order of PSK introduced.

Some actual Intelsat TDMA testing has been carried out with an 8-level PSK system. At the laboratory experimental stage 16-level systems have been examined. The modulation efficiencies are summarized in Table 4.1.

Table 4.1

Method	Modulation Efficiency: Bits s^{-1} Hz^{-1}
2-level PSK	0.5
4-level PSK (QPSK)	Approx. 1 to 1.5
QPSK staggered by ½-bit in time (SQPSK)	Approx. 1.8
8-level PSK	Approx. 2
16-level PSK	4

Clearly QPSK already offers an improvement by a factor of between 2 and 3 over simple 2-level PSK. In order to achieve this amount of improvement yet again, it seems that the 16-level complexity is required. Thus, most existing or shortly contemplated systems are orientated towards 4-level or quaternary PSK (QPSK).

4.3 A Series-Type QPSK Modulator

An arrangement providing this function is shown in Fig. 17.

The carrier source provides a pure sinusoidal signal which is relatively stable and noise-free. This microwave signal is passed through the first and second circulators and, in the absence of the circuitry connected to the intermediate ports, this signal would simply pass on to be amplified by the final stage. In this situation the output to the antenna from the reflection amplifier would just be the unmodulated carrier—at increased power.

The combination of the first and second circulators produces QPSK of the carrier. Examining the first circuit, containing the quarter-wave line, this transforms the short-circuit (S.C.) to an open-circuit at the diode. The remaining length of line from the diode to the circulator is an integral number of half-wavelengths long. Thus, when a '0' is applied to the diode it does not conduct and the signal passing through the circulator suffers zero change of phase. However when a '1' is applied so that the diode does conduct, the signal experiences an effective short-circuit in the plane of the diode and a 180° phase shift is now evident. So this first circuit produces 0°, 180° PSK of the signal.

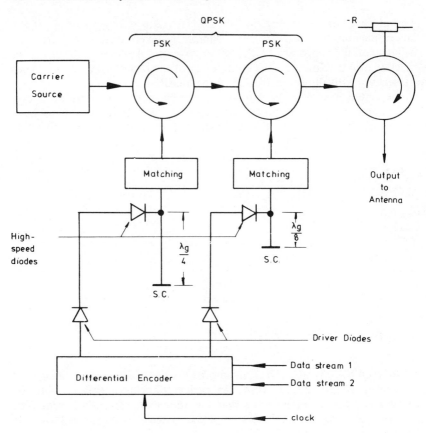

Fig. 17 A series arrangement giving QPSK modulation of a microwave carrier.

The second circuit, using similar principles, also produces PSK of the already 0°, 180° signal. This second circuit provides the essential intermediate 90° phase shift, by means of the eighth-wavelength line. The overall QPSK angles are thus the conventional 0°, 90°, 180° and 270°.

PIN diodes act as effective and reliable high-speed diodes at C-band (4–8 GHz), where final-stage IMPATT-diode reflection amplifiers yield 10–30 W C.W. In the 19–21 and 28–29 GHz bands GaAs Schottky barrier diodes have to be used to cope with the 400 mbit s^{-1} modulation rate. (PIN and Schottky barrier diodes are dealt with in Chapter 7). The wholly digital sub-systems, exemplified by the differential encoder, will not be discussed here since any treatment of such sub-systems belongs elsewhere. Some good papers and articles on digital microwave techniques will be found amongst the references and books listed on page 71, in particular, the article by Dudek and Robinson (*Electronics and Power*, May 1981).

5

Microwave Electron Tubes

In spite of the rapid proliferation in recent years of solid-state active devices there is still a need for the employment of vacuum tube devices in applications where high power levels are involved, such as the output stages of transmitters. At microwave frequencies the changeover point is at mean power levels of 1 to 10 watts or somewhat more and at pulsed power levels of 10 to 100 watts, or somewhat more, since solid-state devices cannot be manufactured with active regions capable of absorbing the energy lost due to inefficient operation at these levels. However, conventional valve devices which basically operate by varying the number of electrons passing through the device by the voltages applied to control grids begin to suffer from the effects of interelectrode capacitance, lead inductance, and electron transit times between electrodes being appreciable fractions of an r.f. cycle at around 100 MHz. By careful design the frequency range of these valves may be extended to around 3 GHz for a triode, but the power output and bandwidth performance degrade rapidly with increasing frequency.

An important selection of microwave valves and tubes is described in this Chapter, ending with a state-of-the-art summary.

5.1 The Planar Triode

The planar construction (Fig. 18):

(a) Provides ideal terminations for coaxial line resonators and thus radiation and lead inductance losses are reduced.

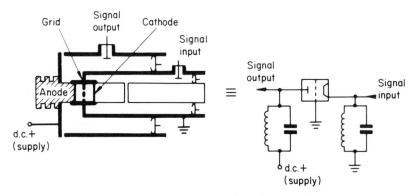

Fig. 18 The construction and equivalent circuit for a planar triode.

(b) Permits the use of small electrode areas which reduces interelectrode capacitances.

(c) Enables the electrodes to be spaced close together to reduce the loss associated with transit-time effects (typical grid – cathode spacing, 0.1 mm).

For efficient operation at high frequencies the cathode current density has to be high (several Acm^{-2}). The concentric line amplifier shown (Fig. 18) may be converted to an oscillator simply by inserting a feedback probe between the two resonators.

5.2 The Klystron

The basis of klystron operation depends on the very phenomenon that reduces the efficiency of conventional valves, i.e. transit time.

The velocity of the electrons constituting the electron beam is modulated by the input signal which is fed to the buncher cavity and this velocity modulation becomes electron density modulation along the drift tube (Fig. 19). The bunched electrons pass the catcher gap when the r.f. field there opposes their motion and hence the r.f. power in the density modulated beam is extracted and fed to the load. If further undriven intermediate cavities are added at bunching planes the bunching is enhanced and increased gain and efficiency is obtained.

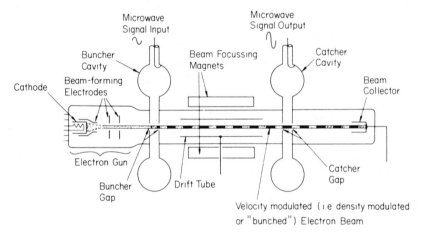

Fig. 19 The two-cavity klystron amplifier.

To maintain a uniform beam cross section down the drift tube either magnetic or electrostatic focussing may be employed.

Multi-cavity klystron amplifiers, employing four or more cavities, produce several kilowatts of r.f. power (C.W.) over bandwidths up to a few hundred megahertz, centred on microwave frequencies. Thus they compete, in *some* applications, with travelling-wave tubes (TWTs) and they also provide the main amplifier in tropospheric-scatter transmitters. In pulsed form they can provide several megawatts of power at the lower microwave frequencies, therefore competing to some extent with magnetron oscillators for radars.

At u.h.f., multi-cavity klystrons represent the principal television transmitter output amplifier.

5.3 The Reflex-Klystron Oscillator

In the reflex-klystron there is only one cavity and no input termination and noise on the electron beam acts as the initial input signal. The electrons become velocity modulated and drift into a region containing a negatively biased repeller electrode where the electrons are turned back towards the original cavity. If the repeller distance and bias are correctly chosen the electrons will be bunched by the time they pass through the cavity gap and energy will be fed into the cavity, some of which will be required to maintain oscillation and the rest will be available for a load.

Although this type of device is only capable of very small fractions of a watt of output power (which could more easily be provided by a Gunn or IMPATT diode) the reflex-klystron can be made to have exceedingly low noise properties. Because of this these devices are still in common use as pump oscillators for parametric amplifiers ('paramps'), which are briefly described in Chapter 6.

5.4 The Travelling-Wave Tube (TWT)

The essential features of a low power travelling wave tube are indicated in Fig. 20. The low level input signal fed in at the input coupler travels along the helix and with suitable spacing between turns the axial component of velocity can be slowed down to match the electron beam velocity. In contrast with the klystron there is a continuous interaction between the signal field travelling along the helix and the beam and thus the process of velocity modulation, density modulation and increased interaction with the helix field gives rise to an exponentially increasing amplitude of signal along the length of the tube (Fig. 21). The extraction of energy from the beam slows it down and hence the initial beam velocity is made slightly greater than the axial component of velocity of the helix field.

For high powers the slow wave structure has to be capable of dissipating more heat and the helix is replaced by a stack of coupled cavities. The device then has a narrower bandwidth and might be considered to be the limiting case of a multi-cavity klystron.

Owing to the bi-directional properties of the slow wave structure and possible mismatch at the input and output terminations some attenuation has to be introduced at a point along the length to prevent oscillation. This might take the form of a graphite coating adjacent to the helix in the low power device or a sever (double taper) with matched terminations in the case of the high power slow wave structure.

As with the klystron the r.f. power output is supplied from the kinetic energy of the beam, but since for the TWT the beam and wave velocities must be similar for power exchange to occur there is an inherent limitation on the tube's efficiency. This is compensated for by the fact that by correct

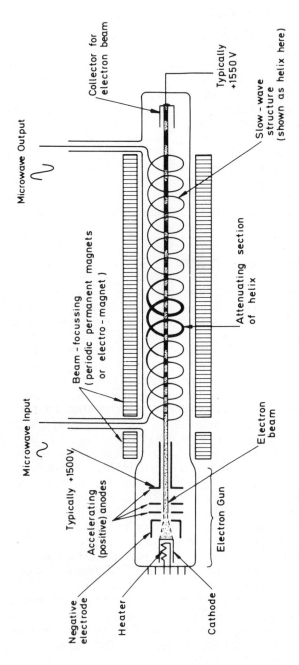

Fig. 20   The basic features of a helix-type travelling-wave tube (TWT).

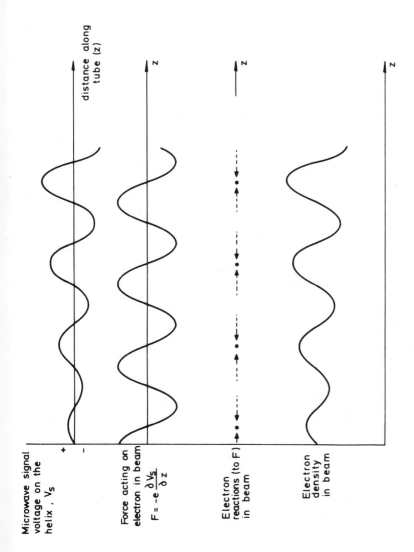

Fig. 21 Signal amplification processes in a TWT.

design of the slow wave structure the wave velocity may be kept constant over frequency ranges of 2:1, and in addition somewhat higher gains can be achieved.

These types of tubes can produce C.W. power amounting to several kilowatts at frequencies up to a few gigahertz (helix-type), or up to 10–30 GHz or more (coupled-cavity type). Pulsed power can approach 100 kW and thus some applications are high-power pulse amplifiers for wide-band frequency-agile scanning radars.

The C.W. devices find military applications commonly in electronic counter-measures (ECM) jammers for noise/deception output stages operating over octave (2:1) frequency bands. Another very significant application is in satellite communication systems ground (Earth) station transmitters, both civil and military. In spread-spectrum systems, a 'hard-limiting' characteristic is desired—which means the maintaining of full or saturated output power over a broad range of input signal conditions. This is achieved by careful design of the r.f. structure of the TWT, the use of multiple attenuators and the cascading of two TWTs with additional equalizers and isolation filters.

5.5 The O-Type Backward-Wave Oscillator (BWO)

This is a development of the TWT amplifier with built-in feedback, where an electron beam delivers energy to a wave travelling backwards along the slow wave structure. The amplified backward-wave velocity modulates the beam and also provides the output energy of the oscillator from a termination at the electron gun end of the tube. The forward wave is absorbed at the collector end by a non-reflecting termination and plays no further part in the action.

These tubes have been largely overtaken by solid-state devices.

5.6 The M-Type Backward-Wave Oscillator (or 'Carcinotron')

This is similar to the O-type except that the slow wave structure and the electron trajectories are circular. It can be operated so that it produces a very high noise output over a wide frequency range and thus is used for microwave jamming. It is also very frequency-agile, it can be 'noise modulated' and typically has a 40% electronic tuning range.

5.7 Crossed-Field Amplifiers (CFA or, more generally, Crossed-Field Tubes: CFT)

The electron stream formation mechanism in a crossed-field tube is quite different from the conventional transparent grid method as used for klystrons and TWT's, in that it relies on the motion produced in an electron under the influence of perpendicular electric and magnetic fields. Rather than an accelerating drift in the direction of the electric field, circular motion perpendicular to the magnetic field is combined with a constant drift perpendicular to both fields to produce a cycloidal motion. The CFT construction is then with just anode and cathode with a magnetic field applied parallel to them producing an average electron drift parallel to the plates. As in the TWT this electron stream is allowed to interact with a travelling r.f. wave, in this case by making the anode a planar version of a slow wave structure.

The presence of the r.f. field, when travelling at the same velocity as the electron drift velocity, upsets the balance of the forces on the electrons and spikes of electrons travel to the anode. As they pass from cathode to anode they interact with the r.f. fields so as to transfer potential energy due to the static electric field into r.f. power, causing amplification of the wave. Since the potential energy available depends only on the electric field, and drift velocity on the ratio of electric and magnetic fields, r.f. power may be extracted from the tube without interfering with the velocity matching of beam and wave, allowing improved efficiency to be achieved. The slow wave structure design is however complicated by the need to dissipate heat generated by the electrons striking it.

The planar form of the CFT is not the commonest found in practice, because advantages of compactness can be obtained by using the circular cylindrical form of the tube. In this version the design of amplifiers must ensure that electron spikes decay away in the zone between output and input to avoid feedback effects. Oscillators may easily be produced in this configuration by deliberately violating this restriction; in the magnetron the slow wave structure is replaced by a set of resonant sections. In this device the rotating 'wheel-spoke' bunches of electrons are similar to those of the amplifier configurations, but the resonant sections have a standing wave electromagnetic field distribution rather than a travelling wave.

Because of its great importance, the magnetron is now considered in its own right.

5.8 The Magnetron

The magnetron consists essentially of a copper ring, into which resonant holes have been machined. These holes connect with the interior of the ring via slots. A cathode is mounted concentrically with the rings and an axial magnetic field is applied to the whole system (usually by permanent magnets). The application of a high voltage to the ring (anode) results in the emission of electrons from the cathode. In travelling towards the anode their paths are curved by the longitudinal magnetic field and their trajectory takes them past the slots cut in the outer ring. If conditions are chosen correctly, the electrons will give up some kinetic energy as they pass the slots. The energy given up will increase the amplitude of the oscillations in the cavity. This situation is illustrated in Fig. 22.

When the magnetron is oscillating the cavities are coupled together by oscillatory E and H fields and energy is therefore extracted from all cavities by a loop within one cavity.

In addition to their basic helical motion the electrons are also velocity modulated and tend to form bunches as they pass the cavities. In an 8-cavity magnetron these bunches form four 'spokes' centred on the cathode and these rotate in synchronism with the oscillatory field.

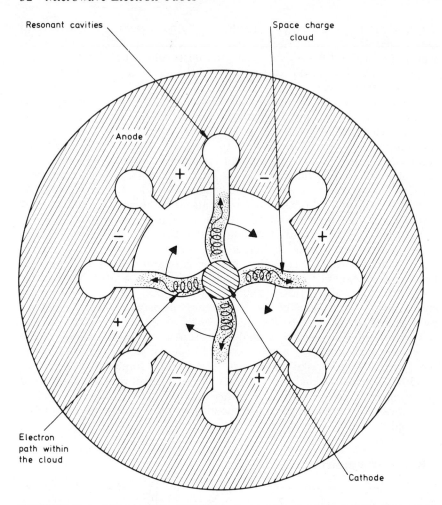

Fig. 22 Electron paths in a cavity magnetron.

The magnetron is capable of producing very high pulses suitable for radar applications, and of (typically) kilowatt continuous power used, for example, in microwave cooking.

Major applications for the magnetron are in radar navigational aids and in the distant detection of aircraft, ships and vehicles.

Typical output powers in such applications are, at a frequency of approximately 10 GHz:

3 kW (peak, pulse), 1.5 W mean,

and 75 kW (peak, pulse), 37 W mean.

Power conversion efficiencies are typically 40–70%.

Since the cathode is subject to electron bombardment it has to be of rugged construction and the heating may be turned down or in some instances off once the device is oscillating at full power.

At lower powers C.W. magnetron oscillators can be made to possess a linear voltage-tuning characteristic (from 20% of operating frequency to as high a range as 3:1 by using a modified form of anode block and different loading conditions. These then find application as a carrier source and receiver local oscillator.

5.9 Summary of Some Significant Tubes

Operating principles and construction are summarized in Table 5.1.

Table 5.1   A Comparison of Operating Principles and Constructions

Device:	Klystron	TWT	CFT
Energy converted:	Kinetic	Kinetic	Potential
R.f. structure:	Resonators	Slow-wave structure	Slow-wave structure
Oscillator form:	Cavity feedback or reflected electron beam	Reversed r.f. wave diection	Standing wave resonant r.f. structure
General operating characteristics:	Narrow band. High power. Moderate gain. Moderate efficiency. Fairly simple construction	Broad band. High power. High gain. Low efficiency. Complex construction.	Broad band. Moderate (magnetron high) power. Low gain. High efficiency. Complex construction.

As with all electronic devices, the choice of a particular unit will depend very much on the application involved. A brief outline of typical applications is now given:

(a) TWT—Commonly used in ECM jammers requiring, typically, greater than 100 watts of power for noise/deception power output stages over wide (octave) bands. The TWT is also used as a high power pulse amplifier for wide-band frequency-agile scanning radars. BWO's are used as signal sources in laboratory test systems, but are tending to suffer competition from solid-state devices below about the 1 watt power output level.

(b) Klystron—High power pulse amplifiers for radars with narrower frequency band requirements with improved efficiency relative to TWTs. Cavity feedback oscillators provide a frequency stable source of microwave power at around 100 watts for communication links and other medium power requirements. Reflex-klystrons are still in common use as 'paramp' pumps in spite of competition from solid-state devices.

(c) CFT—The magnetron is the commonest form of these devices being a source of high peak power pulses (up to 10 Mwatts) with very efficient performance. CFT amplifiers have been used as more efficient output stages in wide band pulse amplification applications for radar in preference to the less efficient TWT.

The *average* power obtainable from several types of device, as a function of frequency, is indicated in Fig. 23.

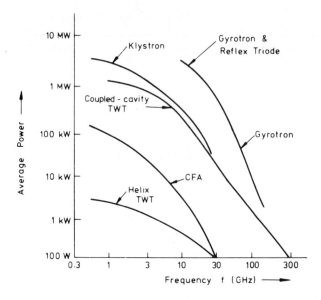

Fig. 23   The power—frequency capabilities of various microwave tubes.

5.10 The Gyrotron and the Reflex Triode

The gyrotron depends upon electrons rotating at a cyclotron frequency, whilst under the influence of a strong magnetic field.

The device was originally developed in the late 1970's for producing very high microwave power at frequencies up to a few *hundred* gigahertz. Several hundred kilowatts (C.W.) has been developed at 60 GHz, and 1000 MW (1 GW) 70 ns pulses have been produced at 10 GHz (USA, USSR, and Culham labs, UK work). An excellent account of the gyrotron has been given by Smith whose article 'The Gyrotron' appeared in *Electronics and Power* in May 1981. (See item 8 under suggestions for Further Reading.)

Under pulsed conditions the *reflex triode* is competitive with the gyrotron, at least at 'X-band' (8-12 GHz). The reflex triode causes an

Table 5.2 Alternatives for 'Ultrahigh' (1 GW pulse) Microwave Power, Below Approximately 12 GHz Frequency

Parameter	Magnetron	Gyrotron	Reflex Triode
Accelerating potential	360 kV	3·3 MV	1 MV
Beam current (kA)	12	80	20
Pulse length (ns)	30	70	20
Axial field	800 G	10 kG	75 G
Power out (GW)	1	1	1
Frequency (approximate) GHz	3	10	10
Efficiency (%)	35	0·4	5

oscillating dipole of electrons to be set up and output energy is extracted from these dipoles. Both the gyrotron and the reflex triode involve relativistic effects in their operation.

A summary of these devices, and advanced magnetrons, is presented in Table 5.2.

A reflex triode has produced 3 GW pulses at 10 GHz, and the US Army, for example, currently has these devices under intensive development for battlefield applications.

6

Amplifiers: A General Comparative Discussion

6.1 Introduction

About 20 years ago the only semiconductor devices likely to be used in microwave circuits and systems were varactor, tunnel and 'crystal' detector diodes. For the same era it is also true to say that circuits operating at much lower frequencies did not enjoy the advantages of integrated circuit design approaches, even transistors were hardly into the megahertz operating ranges.

Just as silicon IC's ('chips') now dominate such a vast portion of the electronics scene so solid-state microwave sub-systems are also increasingly the mainstay of microwave communication systems and radars. Of course, valves or 'tubes' are still necessary for really high power elements in these systems. For example, travelling-wave tubes (TWTs) provide several kilowatts C.W. power in most satellite communications ground-station transmitters. Valves or tubes therefore tend to be found in fairly specialized situations, as described in Chapter 5.

Microwave solid-state amplifiers can be based upon any one of a fairly wide variety of devices:

(i) Microwave transistors, (bipolar, MESFET; see Chapter 3).

(ii) Varactor diodes (for 'parametric' amplifiers).

(iii) MASERs. (Microwave Amplifiers using Stimulated Emission of Radiation).

or

(iv) Gunn, tunnel, IMPATT or BARITT diodes (for 'reflection' amplifiers; see Chapter 4).

Parametric amplifiers ('paramps') and MASERs will be considered only briefly here. Paramps are under quite strong competition from microwave transistor amplifiers for many applications, although they remain important at frequencies beyond about 30 GHz. They (paramps) are also being designed on a 'low-cost' basis at lower microwave frequencies and can be cooled to provide a noise performance surpassed only by MASERs.

MASERs are quantum-electronic devices. They often employ a ruby crystal immersed in liquid helium and are used almost exclusively in radio telescope receivers. (There are also exist ammonia MASERs that are used as standard frequency sources.)

6.2 Selection of a Suitable Type of Amplifier

As with lower-frequency amplifiers, microwave amplifiers fall into three general categories:

(1) Low-noise.
(2) Relatively high gain.
(3) Relatively high output power (usually with specified 'low' distortion and 'tolerable' efficiency).

Some critical questions must therefore be answered before a proper choice of active devices and circuits can be made. Such questions should include:

(a) What are the operating signal frequencies?

(b) Is the sub-system requirement low-noise, substantial gain or substantial output power?

(c) In each case in (b), how low must the noise factor (or temperature) be, how high the gain or how much power?

(d) What is the approximate range of anticipated input signal power?

(e) Is it desirable that the signal, whilst undergoing amplification, be subject to a digital modulation process such as phase-shift-keying?

(f) In what transmission medium, waveguide, coaxial line, strip-line etc., does the input signal arrive ready for amplification?

The answers to questions (a), (b) and (c) especially will determine whether or not a transistor amplifier is even a possible candidate. The situation is, month-by-month, changing—sometimes dramatically. Good review journals should be studied to maintain an up-to-date appraisal of the developments, but be wary of 'research' advances compared with truly commercially available devices. The present (1980's) situation will frequently be referred to here, especially where numerical information must be given.

6.3 Noise Performance Comparisons

The noise temperature T_e of any circuit or device having a noise factor F can be determined using the formula

$$T_e = (F - 1) T_s$$

where T_s is the standard source temperature (290 K, unless, exceptionally, otherwise specified). Microwave engineers often seem somewhat undecided whether to use noise temperature, T_e or noise factor, F, in equivalent situations. The graph of Fig. 24 indicates the order of performance which can be expected from circuits based upon various devices.

Microwave mixers tend to be rather noisy, so these are usually introduced further on in the system. The mixer referred to in the graph would generally be of a diode variety. By using an active device such as a FET to perform the mixing function the advantage of having conversion gain, rather than loss, may be enjoyed. It should be observed that silicon bipolar transistors generally yield a worse noise figure than gallium arsenide FETs. Silicon bipolar transistors also do not stretch so far up in frequency as GaAs FETs. On the other hand silicon technology has many proven advantages compared with GaAs, such as the ability to withstand more thermal stress and reliability. Parametric amplifiers were widely used before the dawn of the microwave transistor era. They are still used in high-sensitivity situations, i.e. where the received signal-to-noise power ratio is small.

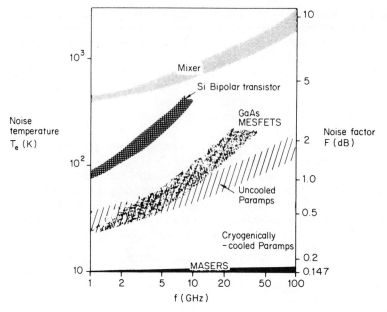

Fig. 24   Noise performance comparisons for a range of devices and circuits encountered in microwave receivers.

Examples are front-ends for:
   (a) Some intercontinental satellite communications receivers.
   (b) Earth receivers for deep-space sources.
   (c) Sensitive radars.
and
   (d) Several systems operating around 30 GHz and upwards—where transistors are not available.

Only the MASER has a noise performance which exceeds that of the best cryogenically-cooled paramp, and MASERs are very bulky and expensive. Paramps suffer at least two specific drawbacks:
   (1) Bandwidths exceeding about 10% are difficult to accommodate.
   (2) They often have to be placed physically close to the receiving antenna.
The trend in the design of complete modern systems, especially communication systems, is to arrange the entire power/noise 'budget' in such a fashion that more economical front-ends such as microwave transistor amplifiers meet the demands quite easily.

At the time of writing (early 1980's) MESFETs have been developed which, at the research phase, will amplify signals at frequencies around 40 GHz under small-signal low-noise conditions. It is considered by some that such devices may eventually reach about 100 GHz capability. If, say, a noise temperature around or less than 50 K is demanded at 10 GHz then, again, transistor amplifiers would be unsuitable (below 2 GHz they could meet this requirement—just).

6.4 A Brief Description of the Parametric Amplifier

Nearly all modern parametric amplifier circuits are coupled using ferrite circulators and have power transferred to the signal by enabling this signal to extract power from a varactor diode. The extra power required is delivered to the varactor diode by means of an oscillator called the 'pump'. An idea of the arrangement is given in Fig. 25.

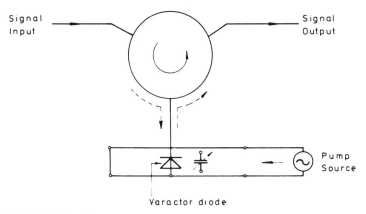

Fig. 25   A simplified schematic diagram showing the main elements of a parametric amplifier.

The varactor diode is maintained under reverse bias and, at any particular frequency f and time t, the energy W stored in the depletion capacitance of the diode will be

$$W = \tfrac{1}{2}\, C\,(f,\, t)\, V^2\,(f,\, t)$$

Notice that both the capacitance and the voltage are functions of frequency and time, making analysis and design complicated.

However, in the simplest terms, energy which has been stored in the depletion capacitance by means of the pump source, is transferred to the signal itself by means of the circulator. Thus net signal gain is obtained. Extra circuitry such as tuning and matching is required for proper operation and the design is critically dependent upon correct choice of varactor diode, as well as the pump frequency (different from the signal frequency) and power.

The most significant feature to note is that:

The only 'active' device in the signal path is the varactor diode.

Therefore the noise performance is only dependent upon thermal noise from the varactor, the circulator, and any noise from the pump source. The latter can be designed for the lowest practicable noise and its output will be well filtered. This explains the superior noise performance attributable to 'paramps' even when uncooled; cryogenic cooling yields a further dramatic noise reduction (see Fig. 24).

By employing suitable tuning, the parametric arrangement can be used to convert the signal to one centred on a lower or a higher frequency. Such circuits are called 'down-converters' or 'up-converters' respectively.

Table 6.1 Comparison of Microwave Amplifiers

Type of Microwave Amplifier		Advantages	Disadvantages
Travelling-wave tube (TWT)		Broad bandwidth High-overload capability Usable to 100 GHz Wide dynamic range High gain and high output power (kW, C.W.)	High power consumption Limited life Warm-up required Large size Large power supply pack
Multi-cavity klystron		High gain and very high output power High efficiency High overload capability	Only moderate bandwidth Limited life Warm-up required Large size Large power supply pack
Travelling-wave maser		Ultra-low noise (a few K) High gain stability Low intermodulation distortion	Large size High price Pump required with intrinsic frequency Intrinsic operating temperature
Reflection Amplifiers	Tunnel diode	High gain per stage Small volume and weight Fairly low noise Octave bandwidth available (usable to hundreds of GHz)	Low output power Burnout protection required Difficult to stabilize
	Gunn diodes	Advantages as for tunnel diode	Fairly low output power (otherwise disadvantages as for tunnel diode)
	IMPATT diodes	High gain per stage Small volume and weight Octave bandwidth Several watts of power Usable to hundreds of GHz	Moderately noisy Low efficiency Difficult to stabilize
Transistors: (bipolar or field-effect devices, especially MESFETs)		Small volume and weight (especially compatible in hybrid form) Broad bandwidth Low price Flat gain response Wide dynamic range High reliability Either low noise, competing with uncooled paramps, or moderate power can be selected	Low gain per stage Frequencies limited currently to about $f < 40$ GHz (low noise) and $f < 12$ GHz (power) Burnout protection required
Monolithic integrated circuits		Ultimate reliability Smallest size Lowest cost — if market demand is sufficient	Noise can be high Less flexibility in circuit design

6.5 Summary of Low-Noise Amplifier Trends

Improved noise performance can be obtained by cooling the device to a fairly modest extent. Thermoelectric (Peltier) cooling is sometimes used. Even uncooled, MESFET amplifiers are currently suitable for earth terminal receiver pre-amplifiers in the 3.7−4.2 GHz band. Similar amplifiers approach the performance required for 'X-band' receivers. Thus, the functions which had been the rightful slots for MASERs in the early 1960's and for parametric amplifiers in the late 1960's and 1970's, are now becoming the province of compact and reliable microwave transistor amplifiers. For direct broadcast satellite (DBS) and domestic satellite (DOMSAT) systems either hybrid or monolithic transistor amplifiers will be used (see Chapter 3).

6.6 A Comparative Summary of Microwave Amplifiers

Table 6.1 gives a fairly extensive comparison of the possible types of amplifiers.

6.7 Noise and Power Gain Trade-Offs in 'Front-End' Amplifiers (Receivers)

One of the most difficult design problems encountered in many communication and radar systems is that of the 'front-end' receiving sub-system. In sensitive receivers this necessarily comprises the antenna, a microwave feed arrangement and an r.f. amplifier. The gain and noise properties of the antenna and the amplifier have a considerable bearing on the performance of the system.

Antenna gain 'G' and system noise temperature T are often combined to yield a figure-of-merit for a receiver—especially in satellite communication systems. The G/T ratio is usually quoted in dB:

$$10 \log \frac{G}{T}$$

(the units are usually quoted as dB even though the argument T of the logarithm is *not* dimensionless).

The system noise temperature has four main contributions:

(i) Atmospheric attenuation in the main beam (about 40%).
(ii) Sidelobe pickup.
(iii) Ohmic losses in the microwave field system.
(iv) The low-noise receiver amplifier.

In this Chapter we are concerned with suitable low-noise microwave receiver amplifiers and such an amplifier invariably has to incorporate at least two stages in cascade, indicated schematically in Fig. 26.

A_1 and A_2 are power gains (as linear ratios) and F_1 and F_2 are noise factors (also as linear ratios).

Clearly the first stage has finite power gain—and it also introduces extra noise into the system. The second stage also has its own power gain and

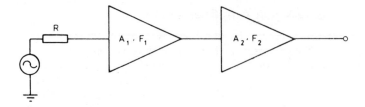

Fig. 26   Two receiver–amplifier stages in cascade.

introduces still more noise into the system. The question is:

> Is there a balanced decision to be made regarding the gains and noise
> factors of the two stages?

To answer this problem we must consider the *overall* noise factor of the two stages.

$$F = F_1 + \frac{F_2 - 1}{A_1}$$

(This is known as Friss' theorem and is derived in many textbooks). This shows that the first amplifier directly affects the noise, whilst the effect of the second amplifier is reduced (especially by gain A_1).

Important points are:
1   The overall noise factor of any cascaded system depends primarily upon the noise factor of the first stage.
2   The effect of subsequent stages is reduced by the power gain up to that particular point in the system.
3   The use of a passive stage, i.e. a stage with a 'gain' of less than 1.0, increases the importance of the noise factor of subsequent stages.

It should be clear that the first stage design must always be concentrated towards a low noise factor (F_1), with 'reasonable' associated gain (A_1). Then the second stage noise factor is less critical to the overall receiver performance and its gain may be made quite large.

$$A_2 \gg A_1$$

As well as system and circuit *choice* and optimization it should be remembered that, where transistors are used, the noise factor also has to be minimized by setting an optimum emitter or source current (d.c.).

6.8 Microwave Solid-State Power Amplifiers: Some Basic Aspects

We have, this far, dealt mainly with low-noise aspects of microwave solid-state amplifiers. The next requirement would usually be (in receivers and also transmitters) for a relatively high gain stage. The achievement of this high gain does not really demand new concepts or particularly elaborate approaches. Problems such as determination of the correct bias conditions and r.f. loading are commonly encountered. Often the amplifying devices can be treated as if they were unilateral and stability is not hard to maintain. (However, in many significant microwave amplifier design problems stability is hard to guarantee.)

It is useful first to examine the power–frequency possibilities offered by some solid-state devices and an abbreviated picture is afforded by Fig. 27.

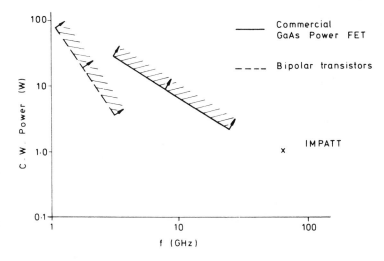

Fig. 27 The power–frequency performance achievable from some representative microwave semiconductor amplifier devices.

This simple outline of the possibilities has one common feature which must be remembered. That is the power–frequency trade-off which always exists for any such devices. The law generally follows a Pf^2 form which is similar to that previously illustrated in Fig. 14.

It is possible to obtain much more power from a complete sub-system simply by using power combiners to add the power from a number of individual devices acting as amplifiers. Microstrip-type power combining circuits can be used successfully in this role and about 10-20 W C.W. at 10-15 GHz is obtainable this way (i.e. suitable for some terrestrial line-of-sight microwave radio relay transmitters, or spacecraft radio transmitters).

Of course, a multi-kilowatt 10 GHz amplifier would require about 1000 devices and an equal number of power combiners—a complex circuit indeed, especially bearing in mind the increasingly serious thermal stresses imposed on successive devices and circuits towards the output end of such an arrangement. Such a circuit is likely to prove more expensive than a vacuum tube at the same power level.

6.9 Some Special Terms used in the Context of Power Amplifiers

Power-added efficiency: η_{PA}

This is defined as follows:

$$\eta_{PA} = \frac{\text{The difference between output and input signal power}}{\text{The } total \text{ electrical input power}}$$

$$= \frac{P_o \text{ (r.f.)} - P_i \text{ (r.f.)}}{P_i \text{ (d.c.)} - P_i \text{ (r.f.)}}$$

If P_i (r.f.) $\ll P_o$ (r.f.) then the efficiency is simply:

$$\eta = \frac{P_o \text{ (r.f.)} - P_i \text{ (r.f.)}}{P_i \text{ (d.c.)} + P_i \text{ (r.f.)}}$$

However, in most power amplifiers the input signal power is still comparable with the output signal power and hence the more accurate η_{PA} definition is important.

Power-added Gain (A_{PA} or 'PAG'):

A definition of this is:

$$A_{PA} = \frac{\text{The extra power 'added' by the amplifier}}{\text{The input power}}$$

or

$$A_{PA} = \frac{P_o - P_i}{P_i} \quad \text{(where all the powers are r.f.)}$$

Note that this gain can never fall below unity. This quantity will frequently be encountered in practical amplifiers.

7

Microwave Passive Diodes

7.1 Introduction: Detector Performance Criteria

There are three main types of microwave detector diode in current use:
  (i) Backward diodes.
  (ii) Schottky diodes.
  (iii) Point-contact diodes.
All three of these will be described in the next three sections. First the ideal detector characteristic is reviewed. This is indicated in Fig. 28.

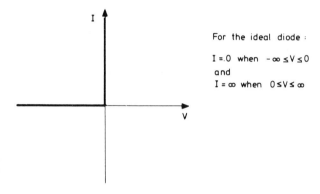

For the ideal diode :

$I = .0$ when $-\infty \le V \le 0$
and
$I = \infty$ when $0 \le V \le \infty$

Fig. 28 Current–voltage characteristic of ideal detector.

If a device with a characteristic like this is connected in a resistive circuit, ideal half-wave rectification is obtained. At high frequencies the charge-storage and parasitic reactances of the device would have to remain negligibly small for the ideal switching action to continue to occur satisfactorily.

A closer approach to practical diode characteristics is furnished by the distortion-free signal rectifier characteristic shown in Fig. 29.

A detector diode is required to produce a d.c. output component as a function of the amplitude of an alternating signal applied to the diode. Through its rectifying action the ideal diode would do this when biased at its operating point, but practical (i.e. non-ideal) diodes, such as those exhibiting a square-law behaviour, $I = aV + bV^2$, also provide detection. It is the bV^2 component that provides the d.c. output from such a detector. Unwanted components are removed by means of a low-pass filter in the detector circuit.

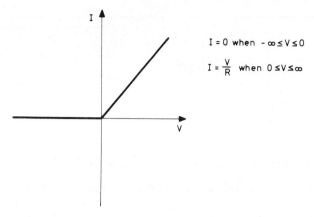

$I = 0$ when $-\infty \leq V \leq 0$

$I = \dfrac{V}{R}$ when $0 \leq V \leq \infty$

Fig. 29 Linear current–voltage characteristic.

With P-N junction diodes, and certain other semiconductor detector diodes, the behaviour is more like an exponential function, i.e. exp (eV/kT). This tends to increase the d.c. output component.

The most efficient and sensitive detector is one which has the sharpest I/V discontinuity at the operating point.

This corresponds to the closest approach to the ideal diode characteristic. Three major criteria are used to indicate quantitatively the detecting ability of a diode and these are:

(i)   Current sensitivity β_0.

(ii) Signal-to-noise ratio *S/N.*

(iii) The so-called tangential signal sensitivity (TSS).

Taking each criterion in turn:

(i) The current sensitivity is defined by

$$\beta_0 = \frac{\text{d.c. output current } I_0}{\text{a.c. signal power absorbed by the diode}}$$

Clearly, the higher this value the better will be the performance of the detector. In practice the r.f. power absorbed is usually of the order of microwatts, and the d.c. output current tends to be a matter of some microamperes. Typical values of current sensitivity for many diodes lie in the range:

$$10 \ \mu\text{A}/\mu\text{W} \leq \beta_0 \leq 20 \ \mu\text{A}/\mu\text{W}$$

(ii) Detector diode noise consists of both white noise and flicker noise components.

Any calculation or measurement of signal-to-noise ratio must take into account the exact circuit in which the detector operates, and the specification of the receiving system involving this diode detector.

(iii) Tangential Signal Sensitivity (TSS).

This is really a subjective quantity to measure. The results obtained depend somewhat upon the judgement of the person making the measurement. Essentially, TSS is a measure of the ability of a detector to discriminate between small signal levels and background noise.

Measurement of TSS is based on the fact that noise, in almost any device, rises as the r.f. drive power rises. Initially, a measurement is made of noise power produced by the detector diode biased as required and connected in the desired circuit; *but with no signal present*. Next, r.f. power is applied to the diode. The level of this r.f. power is adjusted until the output noise power from the diode is just twice the previously observed (no signal) value. This level of r.f. power is measured and its value is termed the Tangential Signal Sensitivity; TSS, and the value is in dBm.

Other methods of TSS 'measurement' are often put forward in the literature, but their results and repeatabilities are often suspect due to the subjective nature of these measurements. The TSS corresponds to a signal-to-noise ratio of approximately 2.5.

It is rather unfortunate that this TSS has become the 'parameter' most commonly used by the industry. Better characterizations are suggested by:

(a) The Minimum Detectable Signal (MDS), which corresponds to a S/N ratio of approximately unity.

or better still;

(b) The Nominal Detectable Signal (NDS), originally suggested by Uhlir, which is actually *defined* for a signal-to-noise ratio of unity.

7.2 The Backward Diode

This is simply a PN junction in which moderately heavy doping has provided a particularly useful I/V characterisic for certain detector applications. If a PN junction device is doped to an extent lying mid-way between a low-voltage zener and a tunnel diode, then the I/V characteristic will also lie somewhere between zener and tunnel characteristics. By correct doping the tunnelling can be kept fairly low over the small forward bias voltage range so that the forward current appears to 'saturate' here at rather a small value. The characteristic takes the form shown in Fig. 30.

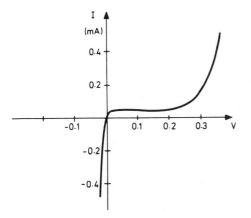

Fig. 30 Current–voltage characteristic for a backward diode.

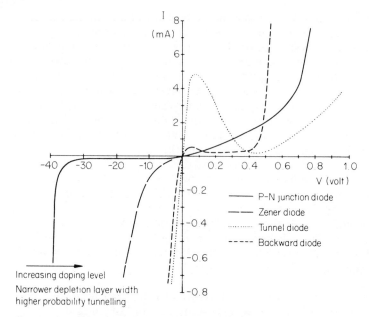

Fig. 31 D.c. characteristics for various types of diode.

This characteristic is compared with that of other diodes in the graph of Fig. 31 showing a family of curves.

The most interesting portion of the backward diode I/V characteristic is the region lying near to the origin. There is a close resemblance to the characteristic required of an ideal diode, if this backward diode curve is considered 'inverted', i.e. 'backward'. Provided the device is operated at low signal levels, such that the rising part of the forward conduction characteristic is never entered, then excellent detection is obtained. The 'leakage' current is typically under 100 μA and the voltage drop (reverse in this case) is generally less than 70 mV.

A major drawback of the backward diode detector is its rather small dynamic range, which makes it virtually of no use for high-drive detectors, or for mixers. However, this type of diode has been used successfully in broadband microwave power-levelling detectors, for example. Particular advantages associated with backward diodes are as follows:

(i) Being fairly heavily doped, the device is rather insensitive to temperature variations.

(ii) The current sensitivity is approximately ten times that of point-contact diodes.

(iii) The TSS is approximately 5 dB better than that for point-contact diodes (typical TSS for a backward diode is about −59 dBm).

7.3 The Schottky-Barrier (or 'Hot Carrier') Diode

The construction of a Schottky-barrier diode is indicated in Fig. 32.

The surface states which occur in the highly-doped N⁺ substrate, in the

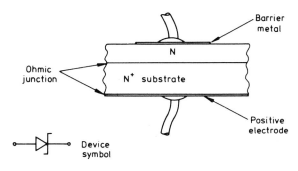

Fig. 32   Schottky diode: structure and symbol.

region in close contact with the 'positive electrode' metal, ensure that only an ohmic effect occurs there. Similar arguments apply to the N$^+$N interface. Therefore no other barriers are produced—the only barrier is between the metal electrode and the N material.

At zero bias a barrier potential exists within the semiconductor in intimate contact with the metal. When forward bias is applied electrons pass through the previously depleted region, reducing the number of positive ions and also reducing the width of the depletion region. If a reverse bias is applied electrons are repelled from the metal region, the depletion effect intensifies and the width of the depletion region increases. Thus it is statistically very much more likely that substantial charge flow will occur under forward bias than under reverse bias. The phenomena just described lead to diode-type behaviour of the metal–semiconductor (or 'Schottky') contact.

The I/V characteristic *equation* looks very similar to that for PN junction diodes, i.e.

$$I = I_o \left\{ \exp\left(\frac{eV}{kT}\right) - 1 \right\}$$

where I_o is the reverse saturation current.

But the actual values of I_o and V, and hence the shape of the curve, are rather different from those applicable for other diodes. The comparisons are summarized in the graph of Fig. 33.

Generally the actual barrier thickness is very much less than the mean free path of the electrons crossing it and therefore collisions in the barrier region may be neglected.

The term 'hot carrier diode' is alternatively used to describe Schottky-barrier diodes because their operation relies upon achieving energetic (i.e. 'hot', $eV \equiv kT$) electrons. It is necessary for the electrons to attain sufficient energy to surmount the potential barrier and become injected into the metal.

These hot carriers take only about 0.1 ps to dissipate their excess energy in the metal. This extremely short time is of no consequence to frequency limitations for the device since the depletion capacitance sets a much lower limit which, however, is typically above 100 GHz. Since, in contrast to PN junction diodes, the Schottky-barrier diode operates with majority carriers

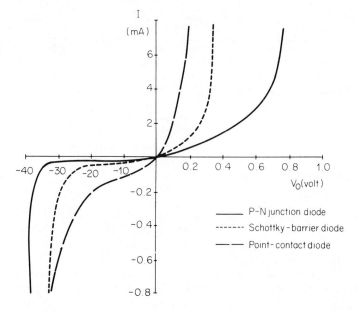

Fig. 33   Comparative d.c. characteristics for PN junction and metal–semiconductor junction diodes.

almost entirely, there is consequently very little charge storage and much higher frequencies can be handled than with PN diodes.

Compared with, for example, point-contact diodes, the Schottky-barrier diode has markedly sharp changes in its resistance around the zero (or near-zero) bias point. This is indicated in Fig. 34.

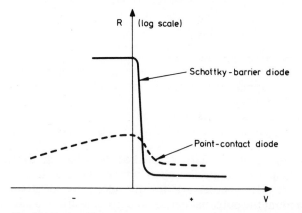

Fig. 34   Resistance as a function of d.c. voltage for metal–semiconductor junction diodes.

The R/V curve applying to Schottky-barrier diodes shows that they offer a closer approximation to the ideal switch than many other diodes. They are therefore preferable, on this basis at least, for detecting, mixing and some switching applications.

These devices also exhibit advantages over many other diodes with their particularly low noise. The main reasons for this are:

(i) The device is unipolar. Only electrons are involved in the conduction processes.

(ii) It is reasonable to operate the device at low d.c. current and voltage bias levels (see the characteristic curves). Because of this the shot noise can be very small. Some Schottky barrier diodes are specified to operate at *zero* bias.

These diodes (Fig. 32) typically have junctions approximately 25 μm in diameter and a 1 μm epitaxial (N) layer thickness of 1Ω-cm resistivity. A short table of appropriate values is given here for the two main materials, silicon and gallium arsenide. In Table 7.1:

N_D = doping density (atoms m^{-3} × 10^{21})
ϕ_n = metal−semiconductor barrier height (V)
W_o = barrier thickness (μm)
C_o = barrier capacitance (pF)
R_s = series resistance (ohms).

Table 7.1 Schottky-Barrier Diodes (Parameters)

	N_D	ϕ_n .	W_o	C_o	R_s
GaAs	1·14	0·71	0·76	0·074	4·84
Si	5·0	0·67	0·32	0·168	13·4

Note particularly the low values of capacitance. The shunt resistance can be quite easily evaluated using a familiar form of expression for semiconductor devices:

$$R \simeq \frac{kT}{e(I + I_o)}$$

At ambient temperature $kT/e \simeq 1/40$, current $I \gg I_o$ usually, and I generally lies within the range 10μA $\leq I \leq$ 200 μA.

These types of devices have many advantages which make them, almost overwhelmingly, strong contenders for high-speed detector applications such as in microwave receivers. The Schottky-barrier structure is also used in transistors where high-speed operation is especially desirable e.g. Schottky TTL for very fast logic circuits, Schottky-gates for microwave field-effect transistors.

7.4 Point-Contact Diodes

This was the earliest form of semiconductor diode ever conceived. Point-contact diodes are still used in some applications, particularly at frequencies beyond about 200 GHz, because modern epitaxial junction devices cannot be satisfactorily manufactured to cope with such high frequencies. These diodes consist of a spring-loaded metal wire pressing against the semiconductor surface. The contact covers a very small area (sub-micrometre) hence the name 'point-contact'. This type of diode is thus really a special case of the Schottky-barrier device and it has been quite widely used because of its low cost and ease of manufacture. The current–voltage relationship is very much like that of P-N diodes.

However, the numerical details of the actual I/V curve differ considerably from those of other diodes and an example is shown amongst the family of comparative curves shown in Fig. 33. The *forward* part of the characteristic is the closest approach to the ideal, but the reverse characteristic is quite poor. A suitable equivalent circuit for a point contact diode would be topologically similar to that already given for the Schottky-barrier diode, but the element values would be different. The capacitance, especially, would generally be less than 0.01 pF. This means that cut-off frequencies of several hundred gigahertz are readily achieved and actual detection at frequencies in excess of 200 GHz is a common application.

In summary, the point-contact diode is a fairly poor detector when compared with the standards set by devices such as Schottky-barrier diodes, but it will sometimes be used in millimetre-wave equipment.

7.5 PIN Diodes

There are several applications in microwave systems where a diode which will switch rapidly 'on' or 'off', in response to an applied control signal, is very useful. PIN diodes provide this facility, as well as yielding a current-variable resistance characteristic which is employed in certain microwave attenuators. A list of applications, which is probably hardly exhaustive, is given:

(i)   Communications channel switching (which may, for example, be pre-programmed).
(ii) Duplexers and multiplexers.
(iii) Phase-shifters (for example in phased-array antenna drives).
(iv) Modulators.
(v)   Radar transmit-receive (TR) circuits.
(vi) Electronically-variable attenuators.

The ideal PIN diode consists of an intrinsic ('I') semiconductor material, having a thickness w_i within the range 10 μm to 100 μm, sandwiched between heavily-doped P and N regions. In practice a true intrinsic layer is difficult to achieve and the 'I' layer is either a lightly-doped high resistivity P region (called a π layer) or N region (called a ν layer). The departures from the ideal are not great and are quite easy to allow for, so the ideal PIN case is considered here.

To facilitate the description the charge, potential and electric field distributions of Fig. 35 are studied.

At zero bias, space charge regions form much as they do in a PN junction device, in the P^+ and N^+ regions adjacent to the I layer—which is completely depleted of free carriers, and therefore is carrying no charge. The electric field rises through each P^+ or N^+ depletion layer, settling to a plateau in the I region. There is only a very small voltage drop across the I region.

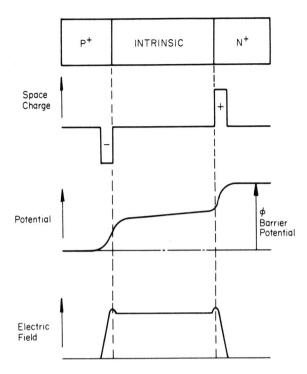

Fig. 35 Pin diode: conditions at zero bias.

In forward bias, there remains a negligibly small voltage drop across the I region and so the drift current is practically zero. If the layer thickness is less than the diffusion lengths the diffusion current will also be negligibly small. Therefore only the recombination current flows under forward bias. The junction capacitance will be mainly due to the I region and will be very small. Under this forward bias carriers are injected from both P⁺ and N⁺ regions and there will be a uniform distribution of these carriers if the I layer thickness is less than the diffusion lengths. A current will flow to maintain continuity in the region due to recombination of the carriers, and it will be equal to the injected charge divided by the carrier lifetime. There is very little un-neutralized charge present in the I region so that the charge densities and carrier lifetimes are essentially identical.

The resistance thus departs from Ohm's law as indicated by the following expression:

$$R_i = V_o I^{-0.87}$$

Often $V_o \simeq 26$ mV and I is in mA units.

This resistance – current law is summarized in the curve of Fig. 36.

It can be seen that, by varying the d.c. forward bias current over a range from 6 μA to 10 mA, the resistance may be adjusted over very nearly three orders of magnitude. This behaviour makes the PIN diode useful in r.f. attenuator and power-levelling applications.

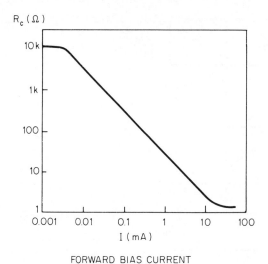

FORWARD BIAS CURRENT

Fig. 36 Pin diode: resistance as a function of forward bias current.

The behaviour under *reverse bias* is now considered. When reverse bias is applied and increased the depletion layers in the P⁺ and N⁺ regions will grow and will cause a small decrease in the total capacitance. The PIN diode behaves, in reverse bias, like an almost loss-free low capacitance in series with a low contact resistance. Its capacitance is somewhat like a varactor, but the actual value of capacitance is much smaller and the capacitance is somewhat less voltage dependent.

Clearly the reverse resistance would be extremely high for an ideal PIN device, but the PνN or PπN practical structure modifies this somewhat. At bias voltages near zero the resistance might be about 10 kΩ ($= R_o$ as shown in the previous forward characteristic graph). As the reverse voltage increases from zero it first serves to clear some free carriers from the ν (or π) region so that the resistance (R_i) rises sharply at first. Once these carriers are cleared the reverse resistance saturates at a MΩ level. The graph of Fig. 37

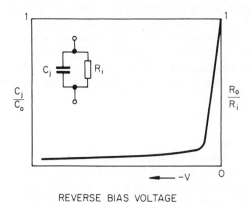

REVERSE BIAS VOLTAGE

Fig. 37 Pin diode: resistance and capacitance variation under reverse bias.

illustrates this behaviour, where the normalized ratio R_o/R_i is used.

Although the reverse resistance is very high at room temperature, it degrades rather drastically as the temperature is increased.

Due to quite long carrier lifetimes in the I region, the reactive and resistive behaviour of the device at microwave frequencies can be calculated using the 'static' picture developed here.

Equivalent circuits appropriate to a PIN diode, for both reverse and forward bias states are shown in Fig. 38.

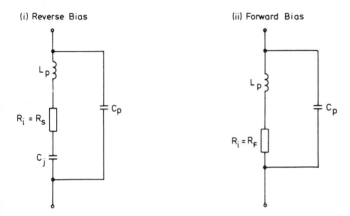

Fig. 38 Equivalent circuits for a PIN diode.

In forward bias R_i is low and dictates the device behaviour; it is shown as R_F here. When reverse bias is applied C_j is much more influential and R_i is too great to have any significant effect. Typical element values for the equivalent circuits, applicable to silicon PIN diodes, are given in Table 7.2.

Table 7.2 PIN Diode Parameters

		Element Values		
		Range	Typical	Units
Wafer parameters	R_s	0·5 to 5	1·5	ohms
	C_j	0·05 to 5	0·5	pF
	R_f	0·1 to 10	1·0	ohms
Package parameters	L_p	0·2 to 2	0·5	nH
	C_p	0·1 to 1·0	0·3	pF

PIN diodes are used, in either series or shunt connection, as switches or phase shifters. Regarding speed and power limitations: PIN diodes that can handle 13 W C.W. (maximum) will switch states in less than 5 ns. Some PIN diodes will operate up to 400 W C.W., but their switching speed has to be traded off to a slower value (typically 1 μs). There are commercially available PIN diodes that handle up to 2 kW pulse power (maximum), considering 1 μs pulses at 10 GHz.

7.6 Step-recovery Diodes (SRD) (or 'Snap-Off' Diodes)

Traditionally, frequency multiplier circuits have employed varactor diodes where the inherently non-linear capacitance/voltage characteristic is utilized. Increasingly this and other applications are being taken over by step-recovery diodes (SRD). An abbreviated list of applications is given:

    (i)   Harmonic generators (frequency multipliers).
    (ii) Impulse generators.
    (iii) Comb generators.

Most diodes, with the exception of the Schottky-barrier diodes, exhibit a charge-storage effect. The influence of this on current waveform response to a sinusoidal voltage drive is illustrated by the comparative waveforms of Fig. 39. All the 'rectifier-type' PN junction diodes exhibit charge-storage

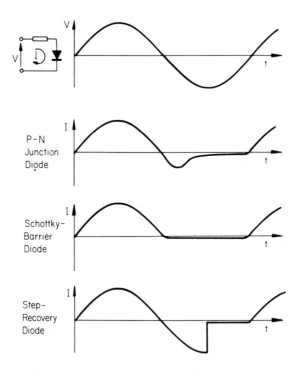

Fig. 39   Charge storage in junction diodes.

due to minority carriers which enter the depletion layer during the forward conduction half-cycle. As a consequence, during the *reverse* voltage half-cycle, a transient current will flow due to the removal of this stored charge by recombination and diffusion out of the region. (This feature is clearly a drawback for a rectifier since zero current flow is, ideally, always desired in the reverse half-cycle). If the frequency of operation is high enough this transient reverse current can occupy a substantial fraction of a half-cycle.

    In the step-recovery diode (SRD) this charge-storage effect is deliberately

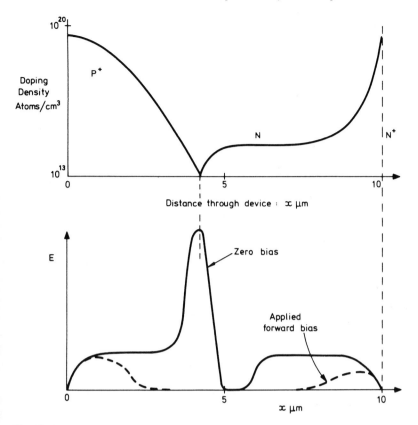

Fig. 40   Doping density and electric field profiles in an SRD.

enhanced by employing a specially designed doping profile which is shown in Fig. 40.

It can be seen that the regions follow a general P⁺N N⁺ sequence, which may seem rather like the PνN sequence referred to when PIN diodes were described. But here, for the SRD, there are two main essential differences: the profiles are carefully controlled and the fairly abrupt changeover from P⁺ to N is ensured. This latter feature provides the localized peak of electric field at the main P⁺ N junction.

We now consider the effects of zero, forward, and reverse bias in turn, starting with zero bias.

Minority carriers are localized around the P⁺ N junction where the high electric field exists. Substantial diffusion current therefore flows through this junction, in fact the high electric field is produced to enable an equal magnitude but opposite direction drift current to flow and cancel the diffusion current. This ensures that equilibrium is preserved.

Next we study the effects of applying a fairly low forward bias voltage. The high electric field at the P⁺ N junction now collapses due to charge carriers being swept through this junction. However, the fields in the P⁺ and N⁺ regions remain substantially unchanged due to the dominant effect of

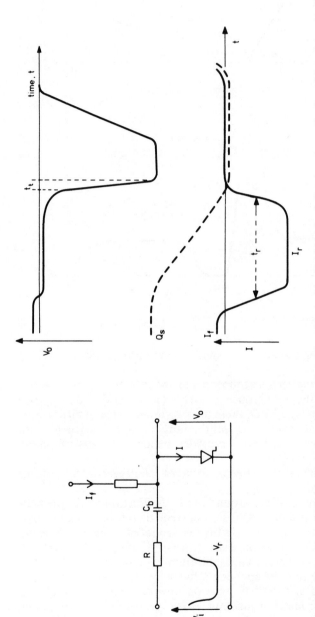

Fig. 41 Pulse-driven SRD: circuit and waveforms (Reproduced from *Physical Models of Semiconductor Devices*, Edward Arnold, 1974 (Fig. 4.13, p. 80) by kind permission of the author, Dr J. E. Carroll).

the high doping levels there. The presence of these fields effectively *contains* (or 'locks in') the previously injected minority carriers within the N region. A charge build-up therefore occurs in this N region and a relatively high capacitance is produced which is associated with this regional charge. These effects can be seen on the electric field versus distance graph.

Finally the effects of reverse bias are considered. The minority carrier current is initially quite high due to removal of the charges which were stored during the previous forward bias application. It should be noted that these charges are stored in a region which includes some of the P⁺ N junction region, where the electric field must increase rapidly again due to the polarity of reverse bias depleting that junction. Thus, the charge removal is comparatively very rapid—typically about 40 ps. The resulting 'snap-back' of reverse current is sharp and is therefore rich in harmonics.

A suitable circuit for an SRD driven by a negative voltage pulse is shown in Fig. 41, where the resultant waveforms are also given.

For this circuit the input pulse is applied through R and the blocking capacitor C_b. Before the pulse is applied, only a d.c. bias current I_f flows and this allows the stored charge Q_s to develop in the SRD. Once the pulse reaches the diode the current swings to a substantial reverse value I_r ($|I_r| > |I_f|$) due to the transient removal of the stored charge Q_s which must now take place. Whilst this current I_r flows the voltage across the device reverses but remains small. At the critical time when all of Q_s has just been completely removed the current will very rapidly fall to nearly zero, i.e. 'snap-back' occurs. This extremely rapid decrease in the magnitude of the reverse current is accompanied by an equally rapid rise in the magnitude of the reverse voltage.

The recovery time, during which charge is removed, is t_r. The transition (or 'snap-back') time is t_t.

In many practical circuits the SRD will be driven by a sinusoidal voltage rather than a pulse. The resultant current waveform, in this case, is shown in Fig. 42.

The processes which occur are identical to those described for the pulsed example and therefore this sinusoidal-drive case will not be described in

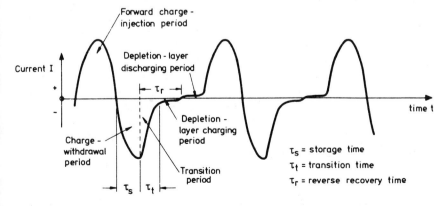

Fig. 42   Current waveform for a sinusoidally-driven SRD.

detail. It can be seen that the time required for the charge to be completely removed equals the storage time (τ_s here). Also, after the fast snap-back transition, the small reverse current remaining charges the depletion layer in the device. When the drive voltage swings positive again, there is a short interval during which the depletion layer is discharged and it is only *after* this that the forward current rises again.

A typical arrangement for an SRD frequency multiplier is shown in Fig. 43. The excitation source is usually a highly-stable crystal-controlled fundamental oscillator. Networks A and B may be tuned circuits as indicated in the simplified circuit diagram of Fig. 44. Here n is the order of multiplication.

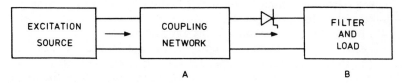

Fig. 43 An SRD-based frequency multiplier.

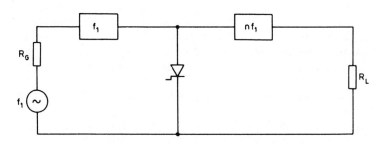

Fig. 44 Tuning arrangements in an SRD-based frequency multiplier.

It is possible, although unusual, for n to be as high as 20; but multiplication factors of 10 are a more common maximum value. If the fundamental source frequency is, say 100 MHz, then at least two separate stages of multiplication are likely to be required where a 10 GHz output is desired. The maximum output frequencies tend to be around 20 GHz, and the power output obtainable can be several watts.

In certain applications a high-speed train of sharp pulses is required; examples occur in radars, comb generators etc. One method of achieving this using an SRD is furnished by the circuit shown in Fig. 45.

The operation of this circuit depends mainly on two features:

(i) in forward conduction the SRD has a very low resistance and behaves almost like a short circuit, and

(ii) in reverse bias the diode exhibits the effective capacitance due to stored charge, until all this charge is removed.

The remaining detailed steps required in the operation, to obtain output pulses, are left as an exercise.

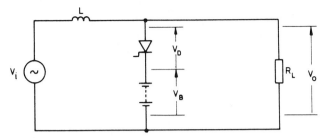

Fig. 45 An SRD-based impulse generator.

7.7 Microwave Varactor Diodes

The basic principles underlying any varactor diode (or 'variable-capacitance', or 'varicap' diode) should be understood from fairly elementary PN junction work. Any PN junction device suffers *depletion* of the charge density local to the interface and the capacitance associated with this depleted zone can be very useful, varying as it does with the applied voltage. There are three main microwave application areas:

(i) In voltage-controlled oscillators (VCOs): f.m. production, automatic frequency control (a.f.c.), phase-locked loops, etc.

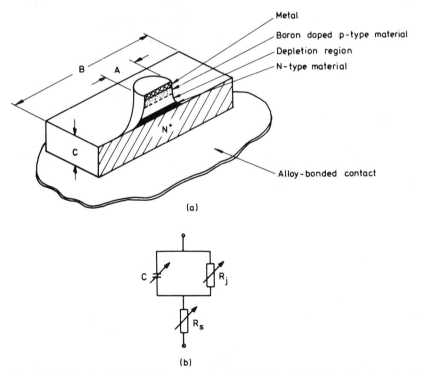

Fig. 46 (a) Section through a silicon varactor diode. (b) Equivalent circuit for the varactor diode chip.

(ii) As the principal component in parametric amplifiers ('paramps') for super-low-noise front-end stages.

(iii) For frequency multipliers (up to three or four times the fundamental), making use of the strong $C(V)$ non-linearity.

Microwave varactors may employ a mesa structure similar to that shown in Fig. 46(a) although some use a Schottky-barrier structure. A suitable (chip) equivalent circuit is depicted in Fig. 46(b). Notice that, in general, *all* the parameters will be bias-dependent. Some representative dimensions and equivalent circuit parameters are shown in Table 7.3 (subscripts 'O' refer to maximum values, and V_B is the reverse-biased breakdown voltage). The table refers to silicon diodes.

Table 7.3 Varactor Diode Parameters

Application	Parameter						
	A (mm)	B (mm)	C (mm)	C_{jO}(pF)	R_{jO}(MΩ)	R_{sO}(Ω)	V_B(V)
Harmonic generator	0·175	0·5	0·15	5·5	>10	0·7	60
Parametric amplifier	0·075	0·35	0·15	2·0	>10	0·45	15

There are three particularly important figures of merit, as follows:

(i) Maximum-to-minimum tuning capacitance ratio (often written C_{TO}/C_T).

(ii) Cut-off frequency.

(iii) Diode Q-factor.

The first quantity, C_{TO}/C_T, is self-explanatory and is especially important with *tuning* varactors (e.g. VCO applications). Values between 8:1 and 10:1 are often obtained for 'hyperabrupt' GaAs devices.

Cut-off frequency and Q-factor are frequently specified on a static basis by the device manufacturer. Then, at some particular reverse bias voltage (V) intended, and referring to Fig. 46(b):

$$f_{CV} = \frac{1}{2\pi R_S C_{jV}}$$

and

$$Q_V = \frac{f_{CV}}{f}$$

where f is the frequency of measurement.

On this (static) basis cut-off frequencies of many hundred gigahertz and Q-factors of several thousand are evident. This is fairly meaningful for tuning varactors, but evades the dynamic question which is important in the case of multiplier or parametric amplifier diodes. For these applications dynamic cut-off frequency and dynamic Q-factor are far better parameters. The details of these are beyond the scope of this treatment.

A summary of the requirements for the various applications is finally given here.

(i) Varactors for Voltage-Controlled Oscillator (VCO) Applications

Here a wide bias voltage range is desired together with a fairly large capacitance range (typically up to about 10:1). A high Q-factor is also important, so that oscillator loading is avoided. Typical circuit applications include:

>   automatic frequency control (a.f.c.),
>   phase-locking,
>   compensation for oscillator frequency chirp during a pulse (e.g.
>   with pulsed Gunn oscillators).

(ii) Parametric Amplifier ('Paramp') Applications

These applications usually present the most severe demands on the varactor diode parameters. All three quantities γ, Q and f_c, should be as high as possible (γ is the capacitance modulation factor). The highest conceivable value of γ approaches 0.5, for a hyperabrupt junction, but this is exceedingly difficult to achieve for parametric diodes. It is interesting that, whilst (static) cut-off frequencies are invariably given by manufacturers *no* data is generally given *at all* regarding Q. There are, however, standard and well-recognized measurement techniques.

Since, for these parametric diodes, only small-signal excursions are involved, the device is usually quite heavily-doped on both sides of the junction and extremely high cut-off frequencies are available. Values as high as 1300 GHz are quoted for commercially available GaAs parametric varactors. (Thus, ultra-low-noise millimetric receivers have been developed).

(iii) Harmonic Generator Applications

Varactor diodes selected for harmonic generation involve quite different considerations to those needed for paramps. The main feature, which will affect conversion efficiency etc., is a relatively large bias voltage range. Operation over a range: $-35\ \text{V} \leq V \leq -125\ \text{V}$ is reasonably typical. Such multiplier circuits are usually much more tolerant of a greatly reduced cut-off frequency and a somewhat lowered Q-factor than those demanded of parametric diodes. Of course, *millimetric* conversion does again require a higher cut-off frequency. (N.B. Step-recovery diodes also find application in frequency multiplier circuits, see Section 7.6).

Problems

1 Explain briefly why Gunn devices are usually made of N-type gallium arsenide, and with wafers of that material only several microns (μm) in thickness. Why would a wafer of *silicon*, having similar thickness, fail to function as a transferred-electron device?

A Gunn diode is used in a coaxial cavity oscillator to produce a microwave output at a frequency of 9.7 GHz. The interval between current pulses is 0.1 ns and the domain velocity is 10^5 ms^{-1}.

Calculate the duration of the current pulses.

2 The product of r.f. output power, load resistance, and square of maximum oscillating frequency is a constant for any transferred-electron device.

Calculate the output power for a 'Gunn' device to operate at a maximum frequency of 70.7 GHz into a 300 Ω load, if another similar device delivers 50 mW C.W. at a maximum frequency of 10 GHz into the same load.

Comment on the likely implications for a short range radar.

3 Calculate the frequency-pulling figure (MHz per ohm) where the constant in the expression referred to in problem 2 is 2000 (W Ω (GHz)^2 units), the output power is 50 mW and the nominal load resistance is 400 Ω. How much will the frequency change if the load resistance becomes 200 Ω?

4 Describe possible techniques for frequency tuning a Gunn oscillator:

    (a) At rapid (μs) rates over several hundred MHz of sweep range.

and (b) More slowly (ms) but over sweep ranges of octave order.

5 In principle, practically any stable microwave source could be realized by following a crystal stabilized v.h.f. oscillator with a high-order frequency multiplier circuit. However special microwave oscillating devices such as Gunn and IMPATT devices have been developed.

Explain this apparent paradox.

6 Describe briefly the constraints which a silicon avalanche diode imposes on suitable d.c. power supply design, and on device design including heat-sinking.

A single-drift (SD) silicon IMPATT diode requires a d.c. supply voltage of 70 V. The maximum electric field, within the device, required for avalanche breakdown is 2.7×10^3 V m^{-1} and the electric field decreases to twice the minimum saturated value at the end of the drift zone.

Calculate:

    (a) The length of the drift zone, and

(b) The natural frequency for device oscillation.

(Assume a linear distance variation of electric field along the drift zone. Take the minimum electric field necessary for saturated conditions as 1.43×10^2 V m^{-1} and the drift velocity at 10^5 ms^{-1}).

7 The product of r.f. power, device reactance, and the square of the operating frequency is a constant for any IMPATT device.

Two devices, one an IMPATT diode and the other a Gunn diode, are required to operate at 10 GHz and they have the following parameters.

	Device capacitance (pF) or resistance (Ω)	Power–frequency–reactance product (Constant) W-ohm-sec^{-2}
IMPATT	0.1 pF	2.4×10^{23}
Gunn	5 Ω	5×10^{20}

Calculate the maximum r.f. power available from each device.

Check the sensibility of your answers by referring to the appropriate sections of the text.

8 An IMPATT diode has a negative resistance versus peak r.f. current characteristic as shown in the figure.

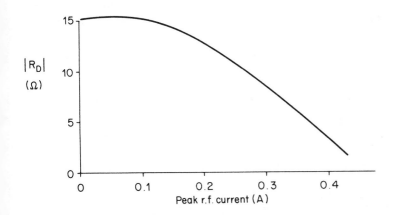

The IMPATT diode has an equivalent circuit consisting of its negative resistance in series with the device's capacitance of 0.2 pF, and the external load may be assumed to be equivalent to resistance and inductance in series. If 0.15 W of microwave output power is required, calculate:

(a) The negative resistance of the device at the appropriate operating point.

(b) The inductance required to tune the device to 12 GHz.

and (c) The efficiency of the oscillator

(If $V_{BIAS} = 70$ V, and $I_{BIAS} = 0.2$A).

Explain briefly the problem associated with the device resistance level and indicate how it may be overcome in practical resonator structures.

9   Approximately the same (2W) output powers are required from two silicon IMPATT devices each of which have essentially similar structures but different capacitances. One device has to operate at 22 GHz and the other at 30 GHz. Calculate:

(a) The capacitance of each device.

and (b) The effective tuning inductances of the cavities in which each IMPATT diode must operate.

How could these inductances be realized in practice?

10   A single silicon IMPATT diode having a capacitance of 0.08 pF normally operates at 15 GHz with an output power of 200 mW. The cavity in which this diode is connected is re-tuned steadily by a few per cent around 15 GHz. Calculate the corresponding output power variation in mW per GHz and mention any assumptions you make.

Estimate the improvement, or otherwise, in this feature if the diode is made of gallium arsenide but in main external respects has the same specifications as the silicon device, i.e. still 200 mW at 15 GHz. (Gallium arsenide has about 30 per cent greater power–frequency–reactance product over silicon).

11   Describe the advantages and drawbacks associated with the use of a magnetic tuning device instead of a varactor diode for tuning a microwave oscillator.

12   Explain the advantages of using a Gunn oscillator, compared with an IMPATT oscillator, as the final microwave locked source in a phase-lock oscillator arrangement.

13   Explain the significance of the GaAs MESFET transistors in microwave systems. Why are these devices capable of such a higher frequency performance than silicon bipolar transistors?

Indicate *two* possible applications of dual-gate GaAs MESFETs.

14   Describe the electrical advantages of connecting a GaAs MESFET chip directly to a hybrid MIC, having microstrip metallization, in the following fashions:

(a) Using bonding wires from the chip to the passive circuitry. These wires may be about 10 μm diameter by 1 mm in length.

(b) Using bonding pads or 'bumps' on the chip and, correspondingly, on the circuit—to facilitate 'flip-chip' attachment.

15   What techniques are available for minimization of noise factor in both microwave bipolar transistors and GaAs MESFETs:

(a) By the device manufacturer?

(b) By the circuit/system designer or user?

16   The first two input stages in a microwave receiver consist of a mixer followed by a high gain amplifier. The mixer has a signal conversion loss of 6 dB and a noise figure also of 6 dB. If the amplifier has a noise figure of 2 dB, calculate the overall noise figure.

17   The system described in problem 16 is changed. Instead of the mixer a silicon bipolar transistor amplifier is used. This has a 2 dB noise figure and

10 dB of power gain. Calculate the overall noise figure in this case.

Compare this system with that of problem 16 and comment on the reasons for the differences in performance. If this system, currently operating at 4 GHz, is required to operate at 11 GHz what alteration must be made?

18   A receiver system has its paraboloidal dish antenna pointing vertically out into space. All the r.f. pre-amplifier elements are secured directly to the feed-point of this antenna (feeder losses may be neglected). Parameters are as follows:

    Antenna effective noise temperature: 5 K
    First amplifier stage power gain: 20 dB
    Second amplifier stage power gain: 10 dB
    Second amplifier stage effective noise temperature: 200 K
    (There are only the two amplifiers).

It is estimated that the overall noise figure is 0.233 dB, to maintain an acceptable signal-to-noise output ratio.

Calculate the noise temperature demanded of the first amplifier stage.

Choose suitable types of amplifiers for each stage and discuss the implications.

19   A solid-state microwave amplifier is supplied by a 10 V d.c. rail. When delivering 1 W of r.f. output power it draws 1 A of direct current from the supply and is driven by 0.2 W of r.f. input power. Calculate the overall efficiency of this amplifier both including and neglecting the input signal power. Comment on the comparison.

20   A range of circulator-coupled reflection amplifiers is to be designed.

Describe and compare the typical anticipated performance of each design, where tunnel diodes, Gunn diodes, and IMPATT diodes are used as alternate active elements.

21   The power gain of any circulator-coupled reflection amplifier is given by:

$$A_{\mathrm{p}} = \left(\frac{|-R| + Z_0}{|-R| - Z_0} \right)^2$$

Where $-R$ is the negative resistance of the device and Z_0 is the characteristic impedance of the connecting lines.

Such an amplifier is to be built in microwave integrated circuit form where the characteristic impedance of the microstrip lines is 50 Ω.

Four active devices are available with the following negative resistances at appropriate bias points:

    (i) $-48\ \Omega$ (nominal).
    (ii) $-1000\ \Omega$.
    (iii) $-1\ \Omega$.
    (iv) $-20\ \Omega$.

Sketch the power gain versus $|-R|$ graph and insert these points.

Explain the implications of considering each device as a possible candidate for use in the reflection amplifier. Hence deduce a general comment regarding the choice of negative resistance, relative to the characteristic impedance.

22 Calculate the appropriate negative resistances and hence discuss the implications for attempting to design circulator-coupled reflection amplifiers with each of the following power gains:
  (a) 33 dB.
  (b) 80 dB.
  (c) 20 dB.

23 Sketch and explain a simple circulator-based arrangement to provide binary PSK modulation of a microwave carrier.

24 The following two microwave links are to be established:
  (a) A tropospheric-scatter system with a 2 GHz carrier and a 20 MHz bandwidth.
  (b) A satellite communication system (ground station) using a 14 GHz carrier and having a 500 MHz bandwidth.
Choose and describe suitable transmitter tubes for each system, giving constructional and operating reasons.

25 Calculate the voltage which must be applied to the final anode of the electron gun in a backward-wave oscillator (BWO), basing this calculation on the equal-velocity approximation (explain this). The helix of this BWO has one turn per mm of its axial length of 200 mm and a mean diameter of 3 mm.

Also determine the average transit time for beam electrons to traverse the length of the tube. How does this time compare with the time per half-wavelength for a 10 GHz signal? Explain.
  (Assume $e/m = 1.76 \times 10^{11}$ C kg^{-1}).
  (Hint: use the kinetic energy/potential energy equality).

26 A multi-cavity klystron amplifier has a power gain of 33 dB and the total supply power (cathode heater, electron gun, focussing, collector) is 5720 W.

Calculate the output power and the overall efficiency of the tube when 0.5 W is input.

27 Suggest, with explanation, a suitable solid-state driver amplifier for the klystron of problem 26.

If this driver amplifier has 17 dBm input power determine its power-added gain (dB) for the situation specified in problem 26.

28 Explain why the extraction of heat should be easier for coupled-cavity travelling-wave tubes than for helix TWTs.

Indicate, with some representative figures, the significance of this feature in high-power TWTs.

29 A *quantitative* description of detector sensitivity, especially one backed by a reliable and meaningful measurement technique, turns out to be difficult to establish. Why is this so?

Explain with the aid of block diagrams, how the tangential signal sensitivity (TSS) might be measured for a microwave detector diode. Show the kinds of instrumentation necessary in this type of measurement and discuss briefly any drawbacks associated with TSS characterization.

30 Give *two* reasons for the particular suitability of Schottky barrier diodes in *microwave* signal detection, compared with PN junction diodes. A suitable equivalent circuit consists of capacitance (C_o), with shunt (R) and series (R_s) resistances.

Explain briefly the physical origin of each parameter.

31 Describe carefully, with the aid of a sketch of a simple sub-system, how PIN diodes may be used to separate the 'transmit' and 'receive' paths in a CW radar system; in particular how adequate protection of the receiver is ensured during the 'transmit' period.

32 Consider the varactor diodes whose parameters were given in Table 7.3.

Calculate the Q-factors for *both* diodes at a frequency of 10 GHz by assuming that the capacitance is one-fifth and series resistance halved in each case at the operating voltage.

Discuss the significance of Q-factor for both the harmonic generator and the parametric amplifier applications.

Answers

1   3 ps
2   1 mW
3   Frequency-pulling figure is -12.15 MHz per Ω; *change* in frequency is -4.14 GHz
6   2.9 μm; 17.24 GHz
7   IMPATT diode 15.1 W, Gunn diode 1 W
8   *One* valid result for part (a) is 13 Ω; 0.88 nH; 1.07%
9   (a) 0.029 pF, 0.0395 pF; (b) 1.8 nH, 0.71 nH
10 5.33 W per MHz (using re-calculated value for product)
16 8 dB
17 2.16 dB
18 9 K
19 9.8%, 10%
21 Power gain as linear ratio: 2401, 1.22, 1.08, 5.44
22 $-52.29\ \Omega$, $-50.00\ \Omega$, $-61.11\ \Omega$
25 2.6 kV (2600 V); 6.67 ns
26 997.6 W; 34.9%
27 9.53 dB
32 41.4, 176.8

Suggestions for Further Reading

There exists a fairly large literature on microwave fundamentals and engineering. What follows amounts to a list, with comments, of particularly useful texts and reference journals.

1 KENNEDY, G., *Electronic Communications Systems*, McGraw-Hill, 1977. There are excellent descriptions of many microwave tubes and semiconductors, although care must be taken because of the pre-1977 date.

2 HA, T. T., *Solid State Microwave Amplifier Design*, John Wiley & Sons, 1981. An advanced treatment of the design of microwave amplifiers, concentrating on transistor amplifiers.

3 LIAO, S. Y., *Microwave Devices and Circuits*, Prentice-Hall Inc., 1980. Recommended for coverage of microwave electron tubes, in considerable depth.

4 EDWARDS, T. C., *Foundations for Microstrip Circuit Design*, John Wiley & Sons, 1981.
Advanced design treatment based upon the principal MIC medium, i.e. microstrip. The design of several actual circuits is developed, including transistor amplifiers. A design/research and development text.

5 Journals which should be seen for new developments:
MSN; MicroWaves; The Microwave Journal.

6 Journals which carry in-depth papers:
IEEE Transactions on Microwave Theory and Techniques (MTT)—the Institute of Electrical and Electronic Engineers, New York, USA.
IEE Proceedings H (Microwaves, Optics & Antennas)—the Institution of Electrical Engineers, London, England.

7 Conferences, the published digests of which are useful:
European Microwave Conference (usually September of each year).
'G-MTT' Symposium (USA, IEEE; usually June of each year).

8 The IEE journal *Electronics and Power*.
In particular the May 1981 special issue on Microwaves is recommended. Two excellent and contrasting papers in this are noted:
PENGELLY, R. S., 'Monolithic Microwave Circuits', pp. 379–384.
SMITH, M. J., 'The Gyrotron', pp. 389–392.

Appendix 1

Decibels (dB): Voltage and Power Ratios

Power ratio:
The basic definition is the following expression:

$$A_p \text{ (dB)} = 10 \log_{10} \left(\frac{P_2}{P_1} \right) \qquad (1.1)$$

Voltage ratio:
Since power is proportional to (voltage)^2, the definition follows from eqn 1.1:

$$A_v \text{ (dB)} = 20 \log_{10} \left(\frac{V_2}{V_1} \right) \qquad (1.2)$$

The powers P_2 or P_1, or voltages V_2 or V_1, may be any values occurring at points of interest in a circuit or system.

Ex 1 An amplifier has an output power of 10 W and an input power of 5 W.
Using eqn 1.1 the power gain is 3.01 dB.
(THIS IS USUALLY APPROXIMATED AS 3 dB AND IS AN IMPORTANT RESULT TO REMEMBER).

Ex 2 An h.f. amplifier in a receiver has an output voltage of 2 V and an input voltage of 1 V.
Using eqn 1.2 the voltage gain is 6.02 dB.

In most microwave work the power ratios are of the greatest significance.
Notice that power gain factors (ratios) such as 10^4 become simply 40 dB.
Attenuators and any other loss-incurring elements result in *negative* dB (check this with $P_2 < P_1$ in eqn 1.1).
Thus, gains and losses through a system are totalled by appropriate additions and subtractions of dB.

dBm
This means dB referred to one milliwatt, and therefore it indicates a specific amount of power. Substitute 1 mW for P_1 in eqn. 1.1:

$$P_2 \text{ (dBm)} = 10 \log_{10} \left(\frac{P_2}{1 \times 10^{-3}} \right) \qquad (1.3)$$

Ex 3
An attenuator delivers an output power of 15 mW. Calculate the power in dBm.

Using eqn 1.3:

$$P_2 \text{ (dBm)} = 10 \log_{10} \left(\frac{15 \times 10^{-3}}{1 \times 10^{-3}} \right)$$

$$= 11.76 \text{ dBm}$$

The use of dBm allows *power levels*, as well as gains and losses in dB, to be determined by additions and subtractions, through the system.

(dBW are similarly defined, referred to one watt).

Automatic calculation with dB is seen to be a relatively straightforward task.

Appendix 2

Radio-Frequency Band Designations

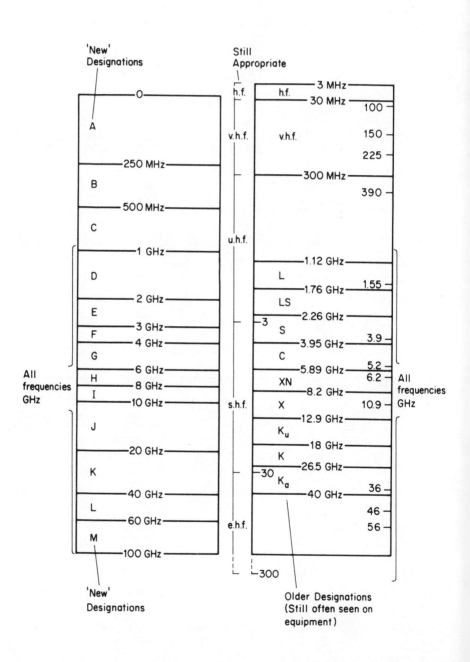

Index

Amplifiers
  comparisons, 40
  crossed-field (CFA), 30
  noise in
    factor, 37–8
    performance comparisons, 37–8
    temperature, 37–8
  power, 42–4
    power-added efficiency, 43–4
    power-added gain, 44
    power combiner, 43
    power–frequency relation, 43
  receivers ('front-end'), 41–2
  reflection, 21–4
  selection, 36–7
Avalanche
  devices, 10–16
  process, 11–12

Backward diode, 47–8
Backward-wave oscillator (BWO), 30
Band designations (r.f. and
  microwave), 74
BARITT diodes, 1, 10
Bipolar transistors, 17, 40

Channel switching, 52
Crossed-field amplifiers (CFA), 30
Crossed-field tubes (CFT), 30–1

Decibel units (dB), 72–3
  voltage and power ratios, 72–3
  dBm, 72–3
  dBW, 73
Detector diodes, 45–52
  current sensitivity, 46
  minimum detectable signal (MDS),
    47
  nominal detectable signal (NDS),
    47
  point-contact, 51–2
  power-levelling, 48
  Schottky-barrier, 48–51
  signal-to-noise ratio, 46
  tangential signal sensitivity (TSS),
    46–7
Digital microwave techniques, 21–4
Direct broadcast satellite (DBS), 41

Diodes
  avalanche, 10–16
  backward, 47–8
  BARITT, 1,10
  detector, 45–52
  Gunn, 2–6, 40
  hot carrier, 48–51
  IMPATT, 10–13, 40
  PIN, 24, 52–5
  point-contact, 51–2
  power-levelling, 48
  step-recovery (SRD), 56–61
  Schottky-barrier, 48–51
  transferred-electron, 2–6
  transit-time, 2, 10–13
  TRAPATT, 13, 15–16
  tunnel, 1
  varactor, 41, 61–3
Duplexers, 52

Electronic countermeasures (ECM),
  30
Electronically-variable attenuators, 52

Ferrite circulator (application), 21–4
Field-Effect transistors, 17–20, 40–41
  gallium arsenide (GaAs MESFET),
    18–20, 36–8, 40–41
  low-noise, 17–20, 37–8, 40–41
  power, 18–20, 40
Frequency multipliers, 7–9, 56–60,
  62–3
Gain (amplifier), 17, 20, 21–2, 29–30,
  33, 37, 40, 41–4
Gallium arsenide, 2–6
  field-effect transistors, 18–20,
    36–8, 40–41
  monolithic microchips, 20, 40
Gunn diodes, 2–6, 40
  domain formation, 2
  oscillator circuits, 6
Gyrotron, 34–5

Harmonic generators, 7–9, 56–60,
  62–3
Hot carrier diodes, 48–51

IMPATT diodes, 10–13, 40
  oscillation condition, 13

Jammers (ECM), 33

Klystrons, 26–7
  multi-cavity (amplifiers), 27, 40
  reflex (oscillator), 27

Magnetron, 31–3
Masers, 36, 38, 40
MESFETs (see Transistors)
Microwave integrated circuit (MIC),
    1, 20
Millimetre-wave equipment, 51–2
Minimum detectable signal (MDS), 47
Modulation efficiency, 23
Monolithic microwave integrated
    circuits (MMICs), 20, 40
Multiplexers, 52

Natural frequency (for TED), 5
Noise
  factor, 37–8
  factor (cascaded circuits), 41–2
  Friss' theorem, 42
  minimization (microwave
      transistors), 19
  performance comparisons, 37–8
  temperature, 37–8
Nominal detectable signal (NDS), 47

Parametric amplifier, 39, 40–41
Phase-lock microwave sources, 7–9
Phase-shifters, 52
Phase-shift-keying (PSK), 22–4
PIN diodes, 24, 52–5
Planar triode, 25–6
Point-contact diodes, 51–2
Power amplifiers, 26–30, 42–4
Power combiner, 43
Power-added efficiency, 43–4
Power-added gain, 44
Power–frequency relation, 43
Power-levelling (diode application),
    48
Power output and efficiency (TED), 7
Power–resistance products, 19
Power transistors, 18–20, 40
Problems and answers, 64–70

Quaternary phase-shift keying
    (QPSK), 23–4

Radar navigational aids, 32
Receivers ('front-end'), 41–2
Reference oscillator, 8

Reflection amplifiers, 21–4
Reflex klystron, 27
Reflex triode, 34–5

Saturation (in TWT), 30
Schottky-barrier (diode), 48–51
Signal-to-noise ratio, 46
'Snap-off' diodes, 56–61
Step-recovery diodes (SRD), 56–61
  comb generator, 60
  frequency multiplier, 60
Synthesis (microwave signals), 7–9

Tangential signal sensitivity (TSS),
    46–7
Temperature (noise), 37–8
Thermoelectric (Peltier) cooling, 41
Threshold field (GaAs), 2
Transferred-electron oscillators
    (TEOs), 2–6
Transistors
  bipolar, 17, 40
  field-effect (GaAs MESFET),
      18–20, 36–8, 40–41
  low-noise, 17–20, 37–8, 40–41
  power, 18–20, 40
Transit-time devices, 2, 10–13
TRAPATT diodes, 13, 15, 16
  pulsed power, 16
Travelling-wave tube (TWT), 27, 30,
    40
  slow-wave structures, 27, 30
Travelling-wave maser, 40
Triode
  planar, 25–6
  reflex, 34–5
Tubes, 25–35
Tunnel diode
  device and applications, 1
  in reflection amplifiers, 22, 40
Two-terminal (active) devices, 1–16

Vacuum tubes, 25–35
  summary, 33
Varactor diodes, 41, 61–3
  cut-off frequency, 62
  parameters, 62
  parametric amplifier application, 39
  Q-factor, 62–3
  summary of applications, 63
Voltage-controlled oscillators (VCOs),
    7, 61–3

Waveguide resonator, 6